Instructor's Manual and TestBank

to accompany

Environmental Science

Creating a Sustainable Future

Sixth Edition

Daniel D. Chiras

Mark Aronson

Scott Community College

Jones and Bartlett Publishers

Sudbury, Massachusetts

Boston Toronto London Singapore

World Headquarters
Jones and Bartlett Publishers
40 Tall Pine Drive
Sudbury, MA 01776
978-443-5000
info@jbpub.com
www.jbpub.com

Jones and Bartlett Publishers Canada
2406 Nikanna Road
Mississauga, ON L5C 3W6
CANADA

Jones and Bartlett Publishers International
Barb House, Barb Mews
London W6 7PA
UK

Cover Image: © Jim Sugar Photography/CORBIS

ISBN: 0-7637-1769-X

Printed in the United States
04 03 02 01   9 8 7 6 5 4 3 2 1

Contents

Foreword

Environmental Science, Sixth Edition presents an overview of issues associated with creating a sustainable future. One of the book's primary objectives is to help students develop an appreciation for the need to think deeply and critically about current environmental issues, as well as to introduce students to skills needed to become an environmental professional, and to focus their attention on constructive methods to effect significant changes in the world that can help lead to a sustainable future.

Environmental science courses can vary widely in their focus and presentation. *Environmental Science*, Sixth Edition can accommodate all audiences. Included below are sample syllabi to assist both the novice and the experienced professor organize a teaching plan. These syllabi may serve as guidelines for your course lectures, or as examples of how to use the text.

In addition to the sample syllabi below, each chapter in the Instructor's Manual contains the following items:

- **Chapter Outline:** An outline of the major headings within each textbook chapter to give an overview of its contents

- **Key Terms:** A complete listing of the boldface, italic, and other important terms in each chapter

- **Chapter Objectives:** General objectives as related to the main headings of each chapter

- **Lecture Outline:** A detailed, sequential outline of each chapter, presented to aid the instructor in preparing lecture notes

- **Suggestions for Presenting the Chapter:** Helpful suggestions and hints for presenting the material in each chapter. Also contains a list of web sites related to topics covered in the chapter.

- **Test Questions:** Each chapter contains up to eighty questions to aid the instructor in preparing quizzes and examinations. These multiple choice, true/false, and fill-in-the-blank questions can be used as they are, combined, or modified to suit each instructor's use. These questions can also serve as the basis for essays and class discussions.

Sample Syllabi

Syllabus 1

I.   Sustainability
-   What is it? – An Overview

II. Principles of Ecology
-   Biodiversity
-   Endangered Species

III. The Human Population
-   Population Growth
-   Urbanization
-   Resource Usage and Distribution

IV. Water Resources
-   The Chemistry of Water
-   Wastewater/Drinking Water
-   Water Pollution and Treatment

V.   Solid Waste
-   Recycling/Remediation
-   Waste Disposal

VI. Air Pollution
-   Noise Pollution

VII. Energy Resources
-   Nonrenewable Energy
-   Renewable Energy
-   Energy Efficiency

VIII. The Impact on the Earth
-   Global Warming
-   Ozone Depletion
-   Acid Rain

IX. Environmental Management
-   Sustainable Solutions

Syllabus 2

I.   Ecosystems
-   Global
-   Regional
-   Local
-   Functionality
    o   Energy and Nutrient Cycling
-   Population Ecology

II. Water Systems
-   Wastewater
-   Pollution

III. Air Pollution

IV. Toxic Chemicals

V.   Human Population Ecology and the Human Impact
-   Population Growth
-   Energy Consumption
-   Food, Agriculture and Soil Resources
-   Urban Growth
-   Pest Control

VI. Public Policy, Biodiversity, and Sustainable Ecosystems
-   Preserving Nature
-   Impact and Planning

Syllabus 3

I. The Earth as a System
- Biogeochemical Cycles
- Energy Transfer
- Human Populations
- Biodiversity
- Ecosystems

II. Maintaining and Preserving Our Resources
- Human Resources
  o Food/Agriculture
- Biological Resources
  o Habitats
  o Endangered Species
- Energy Resources
  o Nonrenewable Energy
  o Renewable Energy
- Physical Resources
  o Air
  o Water
  o Soil
  o Minerals

III. Environment and Our Society
- Urban Development
- Economics
- Waste Disposal – Recycling and Remediation
- Impact and Planning
- Integrating Values and Knowledge

Syllabus 4

I. Environmental Ethics
- Why should we care?
- Why is it important?

II. Air, Climate and Weather
- Air pollution
  o How Pollution Affects Our Climate
    - Acid Rain
    - Global Warming

III. Biomes
- Population Dynamics
- Human Ecology

IV. Soil Resources
- Agriculture
- Land Use

V. Restoration Ecology

VI. Sustainable Solutions

Instructor's Manual
and TestBank

to accompany

Environmental Science
Creating a Sustainable Future
Sixth Edition

Daniel D. Chiras

1

Living Sustainably on the Earth

Chapter Outline

Encouraging Signs on the Path to a Sustainable Society
    Netherlands' Green Plan
    U.S. EPA Green Lights Program

Environmental Protection and Sustainable Development
    What is sustainable development?
    What are the stages of environmental protection?

Sustainable Development: Meeting Human Needs Without Bankrupting the Earth
    The Principles of Sustainable Development
    Personal Actions for a Sustainable Future
    What is the difference between growth and development?

Key Terms

sustainable development	intergenerational equity
life cycle management	intragenerational equity
energy efficiency	ecological justice
pollution control devices	root causes
end-of-pipe controls	biological infrastructure
physical infrastructure	renewable
nonrenewable	carrying capacity
economic growth	

Objectives

1. Discuss the environmental significance of Netherlands' Green Plan.
2. Discuss the meaning of sustainability as it pertains to the environment.
3. Compare the terms "sustainable development" and "economic growth."
4. Compare "end-of-pipe controls" to solutions that address pollution by focusing on "root causes."

5. Define "carrying capacity" and discuss its environmental importance.
6. Discuss the ecological principles of sustainable development.
7. Discuss the social/ethical principles of sustainable development.
8. Discuss the political principles of sustainable development.
9. Suggest some personal activities that are part of a sustainable lifestyle.
10. Define "renewable and nonrenewable resources".
11. Discuss the four stages of environmental protection.
12. Describe the current state of environmental protection in the United States.

Lecture Outline

I. Encouraging Signs on the Path to a Sustainable Society
   A. Many nations are taking steps to slow population growth and create a more environmentally friendly form of commerce.
      1. Netherlands' Green Plan
         a. This plan relies on four basic principles:
            i. Life cycle management - This makes companies responsible for products throughout their life cycle.
            ii. Energy efficiency - Companies can save huge sums of money by using energy more efficiently.
            iii. Sustainable technologies - The government has started a program to develop sustainable technologies and make existing technologies more available.
            iv. Public awareness - The government has publicized this program with the slogan "A better environment begins with you ".
         b. Industries are encouraged to voluntarily eliminate hazardous wastes and pollution without the need of regulation and fines.
         c. The government works actively with industry to identify major environmental problems and establish timetables to reduce pollution. Industries enter an agreement to meet established goals and are fined only if they fail to meet them.
         d. The government allows companies to select the most cost effective technologies to reduce pollution.
         e. Financial and technical assistance is provided by the government where necessary.
      2. U.S. Environmental Protection Agency Green Lights Program
         a. Companies are encouraged to work voluntarily with the EPA to reduce hazardous wastes and energy use.
         b. Companies realize environmental policies that encourage efficient use of resources are also cost effective.

II. Environmental Protection and Sustainable Development
   A. Environmental protection is being incorporated into a strategy called sustainable development.
      1. Sustainable development is a strategy to allow development to occur without destroying the environment.
   B. Environmental protection has evolved from piecemeal local efforts to a much more comprehensive, environmental protection are summarized below:

1. Stage 1: Fragmented, symptoms-oriented approach. This phase began in the 1950's and involved regulations formulated by individual cities and towns. Most of the pollution regulation focused on water pollution, i.e., rivers and lakes polluted by sewage or industrial wastes. Regulations mandated that pollution control devices be used (end-of-pipe controls).
2. Stage 2: Comprehensive, symptoms-oriented approach. This stage has been characterized by nationwide (Federal) regulations that have established uniform minimum standards for environmental quality. States are encouraged to undertake environmental protection and in many cases are required to enact regulations and provide enforcement. This stage began in the 1960's and continues today . The main focus still remains, however, on end-of-pipe controls.
3. Stage 3: Fragmented, root-level approach. A handful of nations are entering a new stage of environmental protection where efforts examine the root causes of environmental problems. This phase is characterized by the restructuring of human systems for sustainablility based on sound scientific and ecological principles.
4. Stage 4: Comprehensive, root-level systems approach. This stage would involve all levels of society: local, state, national and international communities. Restructuring basic human systems would be emphasized including: energy, industry, agriculture, transportation, housing , waste management and water supplies. This means environmental protection would be integrated into every decision we make. We must pass into this stage if we are to have a sustainable future.

III. Sustainable Development: Meeting Human Needs Without Bankrupting the Earth
   A. Sustainable development is a means of meeting present needs in ways that do not impair future generations and other species from meeting their needs.
      1. Sustainable means to be able to keep in existence, to maintain and endure.
      2. Development is a step or stage in advancement or improvement.
      3. Sustainable development is a way of improving or advancing our culture in a way that can be maintained over the long haul.
      4. Intergenerational equity is an important principle of sustainable development. Intergenerational equity is the responsibility of each generation to ensure that the next one receives undiminished natural resources and economic opportunity. All generations are stakeholders in the planet's well being.
   B. Sustainable development requires strategies that seek to optimize social, economic and environmental conditions simultaneously.
   C. The principles of sustainable development are:
      1. Ecological Principles
         a. Dependence - Humans are dependent on a clean, healthy environment. All resources have their origins in the earth. Our personal and economic well-being depends on a healthy environment.
         b. Biophysical Limits - Real limits exist in the planet's ability to supply resources and absorb wastes from human civilization.
         c. Living Within Carrying Capacities - Living sustainably on the Earth will require steps to live within the Earth's biophysical limits.
         d. Interdependence - The fate of the environment is in our hands. Our behavior will profoundly affect the natural environment in ways that could have serious consequences for all livings things, including ourselves.

2. Social/Ethical Principles
   a. Intergenerational Equity - Present generations have an obligation to meet their needs in ways that do not foreclose on future generations.
   b. Intragenerational Equity - Present generations also have an obligation to act in ways that do not impair others who are alive today from meeting their needs.
   c. Ecological Justice - Human actions should not endanger other species which also have an inherent right to the resources they need to survive.
3. Political Principles
   a. Participation - Building a sustainable future will require unprecedented participation from all sectors of society.
   b. Cooperation - Creating a sustainable society will require cooperation among many different participants.
   c. Addressing the Root Causes - To create a sustainable society, we must focus on strategies that address the root causes of pressing social, economic and environmental problems.
D. Personal actions that support sustainability are:
1. Be frugal. Use only what you need. Be a conscientious consumer.
2. Be efficient. Support legislation and nonprofit organizations that promote energy efficiency and use all resources in your daily life efficiently.
3. Be a recycler. Support legislation and nonprofit organizations that promote recycling. Recycle all wastes that you can and buy products from recycled materials.
4. Support renewable resource use. Support legislation and nonprofit organizations that promote renewable energy. Use renewable resources wherever possible.
5. Help restore the environment. Support restoration projects and organizations involved in restoration ecology.
6. Help control population growth. Support organizations that promote population stabilization and growth management strategies.
E. Growth and development are different concepts.
1. Growth results in an increase in material production and consumption that is likely unsustainable. Economic growth typically expands the infrastructure (roads, buildings, power plants, water projects, etc.) and requires exploitation of natural resources and energy consumption.
2. Development is a strategy that calls for improvements in culture that do not necessarily require further increases in resource consumption, pollution and environmental destruction. Development, as used in this textbook, implies an increase in the quality of our lives and the condition of the environment.

Suggestions for Presenting the Chapter

- Class discussion can help students grasp the concept of sustainability. Breaking the class into small discussion groups to list the activities in their lifestyle that are sustainable or could be made sustainable is an interesting activity.
- The instructor might provide local examples of sustainable development and even provide a field trip for this purpose. Examples might include a recycling center, compost facility, solar building or an industry involved in sustainable activity.
- Local environmental problems provide excellent opportunities for class analysis and discussion. The class can be asked to identify end-of-pipe pollution controls and suggest the

"root cause" of the particular problem. Potential solutions might be discussed in light of the principles of sustainability provided in the text readings.

- Instructors might encourage activism by initiating a sustainable project with the class such as recycling or environmental restoration. Student enthusiasm for such projects is usually contagious and spawns interest in more sustainable activities.
- Asking students to analyze their use of resources at home is an effective technique to help students examine their own lifestyles and behaviors. This could take the form of an energy analysis, recycling/waste stream analysis or might focus on consumptive behavior.
- Recommended web sites:

    Alliance for Global Sustainability: http://www.global-alliance.org/
    EPA Green Lights Program: http://es.epa.gov/partners/green/g-lights.html
    Global Vision: http://www.igc.org/glencree/index2.html
    International Institute for Sustainable Development: http://iisd1.iisd.ca/
    Sustainable Communities Project: http://efn.org/~fodor/pter

True/False Questions

1. ___ Oil is a renewable natural resource.
2. ___ The right of other species to have a clean environment is called "ecological justice".
3. ___ The EPA is the federal agency charged with writing many environmental regulations in the United States.
4. ___ The "Green Lights Program" is part of Holland's environmental plan.
5. ___ Photovoltaic panels are used to convert sunlight into useable electricity.
6. ___ The use of renewable resources is a part of sustainable economies.
7. ___ Wind power is a nonrenewable resource.
8. ___ Living within the carrying capacity of the environment is a principle of sustainable development.
9. ___ The earth's natural systems are the biological infrastructure of modern society.
10. ___ Recycling is a personal action that is a part of a sustainable lifestyle.
11. ___ Fossil fuels are the most sustainable fuel sources that exist.
12. ___ Population growth is not a factor related to environmental degradation.
13. ___ Economic growth is an expansion of human economic activity.
14. ___ Unlimited economic growth is possible since humans are not limited by the "carrying capacity" of the environment.
15. ___ The world will have 250,000 births in the next 24 hours.
16. ___ Sustainable development is a strategy for allowing progress to occur without destroying the environment.
17. ___ The use of smokestack scrubbers to remove harmful gases from industrial emissions is an example of "end-of-pipe" controls.
18. ___ All environmental problems have their origin in unsustainable human systems.
19. ___ Renewable and nonrenewable resources have very real limits.
20. ___ Living responsibly to maintain the environment for future citizens is called "intergenerational equity".

True/False Key:
1. F 2. T 3. T 4. F 5. T 6. T 7. F 8. T 9. T 10. T
11. F 12. F 13. T 14. F 15. T 16. T 17. T 18. T 19. T 20. T

Fill-in-the-Blank Questions

1. _____ _____ calls on us to act in ways that satisfy our needs while safeguarding the welfare of all who are alive today.
2. The Earth is the rightful property of all species according to the principle of _____ _____.
3. Living sustainably will allow the environment to be healthy for future generations. This is the principle of _____ _____.
4. Using pollution control devices is an example of ___ __ ___ controls.
5. The U.S. EPA program to reduce energy consumption by business and government agencies is called the _____ _____ program.
6. _____ _____ is a means of advancing human civilization while protecting and even enhancing the environment.
7. The EPA's program to reduce hazardous waste by industries is called the ____ program.
8. Humans are _____ on a clean, healthy environment for many good and services vital to our personal and economic well-being.
9. Roads, dams, highways and buildings form the _____ _____ of human society.
10. Soil, forests and water are examples of _____ resources.
11. The ability of the land to support a certain population of living things is called the _____ _____.
12. The destruction of the ozone layer by emission of chlorofluorocarbons is an example of humans exceeding the _____ _____ of the earth.
13. The environmental protection efforts of the 1950's, characterized by local government instituting laws required pollution control devices is called the _____ _____ approach.
14. The passage of the major federal pollution legislation during the 1960s was the start of a _____ _____ approach to environmental protection.
15. _____ is a political principle of sustainable development that requires all sectors of society to achieve environmental goals.
16. The root cause of air pollution is the use of _____ ____.
17. Infinite growth within a finite system is _____.
18. Efforts that focus on improving the quality of our lives without requiring continued resource extraction is called _____.
19. _____ _____ can lead to high levels of ultraviolet radiation and may cause skin cancer.
20. _____ are chemicals used as refrigerants and are linked with destruction of the ozone layer.

Fill-in Answers:
1. intragenerational equity
2. ecological justice
3. intergenerational equity
4. end-of-pipe
5. Green Lights
6. sustainable development
7. 33/50
8. dependent
9. physical infrastructure
10. renewable
11. carrying capacity

12. biophysical limits
13. fragmented, symptoms-oriented
14. comprehensive, symptoms-oriented
15. participation
16. fossil fuels
17. unsustainable
18. development
19. ozone depletion
20. Chlorofluorocarbons

Multiple Choice Questions

1. The government agency in the United States charged with writing environmental regulations is the:
   a. USDA
   b. FDA
 * c. EPA
   d. IRS
   e. NOAA

2. The Green Lights program will save over $100 million in five years by reducing:
   a. hazardous waste
   b. fuel efficiency
 * c. electrical energy consumption
   d. economic development
   e. physical infrastructure

3. The use of pollution control devices to reduce pollution is also called:
 * a. end-of-pipe controls
   b. life cycle management
   c. recycling
   d. solid waste management
   e. best available technology

4. Which of the following is **not** a characteristic that makes human systems unsustainable?
   a. use of fossil fuels
   b. low efficiency
 * c. use of renewable resources
   d. exceeding local carrying capacity
   e. rampant consumerism

5. The Netherlands' has adopted a unique environmental policy which reduces pollution by working closely with industry called the

   _____.
   a. Green Lights program
 * b. Green Plan
   c. Green Revolution
   d. Green Party
   e. Root-level approach

6. Life cycle management is a strategy that encourages industry to develop products that can be easily:
   a. thrown away
 * b. recycled
   c. shipped
   d. stored for long periods
   e. produced without energy consumption

7. The first stage of environmental protection started in the 1950's with the use of end-of pipe pollution controls that were mandated by individual cities and towns. This stage is best characterized as the:
 * a. fragmented, symptoms oriented approach
   b. comprehensive, symptoms oriented approach
   c. fragmented, root-level systems approach
   d. comprehensive, root-level systems approach
   e. economic development approach

8. In the past two decades vehicle miles traveled in the United States have increased by:
   a. 10%
   b. 50%
   c. 100%
   d. 150%
* e. 200%

9. Which of the following strategies allows progress without destruction of the environment?
   a. economic development
* b. sustainable development
   c. preservation
   d. biodiversity
   e. ecological justice

10. During the next 24 hours how many square kilometers of tropical forest will be leveled for development?
   a. 180
   b. 280
   c. 480
   d. 680
* e. 880

11. One of the major problems with fossil fuels is air pollution. Which of the following responses is a "root cause" solution to air pollution by fossil fuels?
   a. change from oil to natural gas
   b. add catalytic converters to cars
   c. convert diesel engines to gasoline
* d. convert from coal to solar power
   e. install smokestack scrubbers

12. The Netherlands' Green Plan calls for a reduction of pollution by 70-90% and a doubling of economic output by the year:
   a. 1998   d. 2005
   b. 1999   *e. 2010
   c. 2000

13. Which state has established 272 indicators of environmental health and has set impressive goals for environmental improvement?
   a. Connecticut
   b. California
   c. Colorado
   d. Utah
* e. Oregon

14. During the next 24 hours how many people will be added to the world's population?
   a. 10,000
   b. 25,000
   c. 50,000
   d. 100,000
* e. 250,000

15. Soil, forests and grasslands would be considered to be _____ resources.
   a. recycled
   b. economic
* c. renewable
   d. nonrenewable
   e. physical

16. Which of the following factors is **not** a "root cause" of unsustainable human activity?
* a. recycling
   b. population growth
   c. increasing economic output
   d. use of nonrenewable resources
   e. exceeding the carrying capacity

17. Which of the following are ecological principles of sustainable development?
   a. dependence
   b. biophysical limits
   c. living within carrying capacities
   d. interdependence
* e. all of the above

18. Which of the following terms best describes human reliance on the environment to provide goods and services for human economies?
   a. biophysical limits
* b. dependence
   c. living within carrying capacities
   d. interdependence
   e. participation

19. The Boeing Corporation, one of the largest manufacturers of commercial aircraft, will save over $100 million in five years by reducing energy consumption under which of the following programs?
* a. EPA Green Lights program
   b. NEPP2
   c. Green Plan
   d. EPA Energy Star program
   e. EPA 33/50

20. _____ _____ refers to the responsibility of each generation to ensure that the next one receives undiminished resources and economic opportunity.
   a. Intragenerational equity
 * b. Intergenerational equity
   c. Sustainable development
   d. Biological infrastructure
   e. Physical infrastructure

21. Iron ore, coal and oil would be classified as:
   a. renewable resources
 * b. nonrenewable resources
   c. biological infrastructure
   d. water pollutants
   e. hazardous wastes

22. Acting in ways that honor the environmental and economic rights of people alive today is best described as:
   a. economic fairness
   b. ecological diversity
   c. economic development
   d. intergenerational equity
 * e. intragenerational equity

23. The concept that human actions should not endanger other species inherent right to resources they need to survive is called:
   a. intergenerational equity
   b. intragenerational equity
 * c. ecological justice
   d. ecological diversity
   e. economic fairness

24. The earth's ability to support a limited number of organisms is called the:
 * a. carrying capacity
   b. biological infrastructure
   c. ecological limit
   d. greenhouse effect
   e. competitive exclusion principle

25. The concept that the earth's renewable and nonrenewable resources have limits is called:
   a. carrying capacity
   b. maximum sustainable yield
 * c. biophysical limits
   d. exponential growth
   e. dual bottom line

26. Which of the following responses is **not** a sustainable personal action?
   a. being frugal
   b. being efficient
   c. recycling
   d. supporting restoration projects
 * e. encouraging fossil fuel use

27. Which of the following principles is **not** part of Netherlands' Green Plan?
   a. life cycle management
   b. energy efficiency
   c. sustainable technologies
   d. increasing public awareness
 * e. end-of-pipe pollution controls

28. The World Commission on Environment and Development (WCED) was created by the:
   a. EPA
   b. USDA
   c. World Bank
 * d. United Nations
   e. Netherlands'

29. The earth's natural systems or ecosystems are described by economists as:
   a. physical infrastructure
 * b. biological infrastructure
   c. nonrenewable resources
   d. material outputs
   e. capital gains

30. The buildings, roads, ports and dams of a country form the:
   a. infra-infrastructure
   b. biological infrastructure
 * c. physical infrastructure
   d. physical carrying capacity
   e. industrial infrastructure

31. Regional agreements between nations, such as multilateral treaties designed to clean up the Baltic sea, exemplify which of the following principles of sustainable development?
   a. intergenerational equity
   b. intragenerational equity
   c. ecological justice
 * d. cooperation
   e. biophysical limits

32. Which of the following activities would be sustainable participation in society?
    a. buying "green" products
    b. reducing consumption
    c. recycling
    d. using fuel efficient cars
  * e. all of the above

33. The conservation and reuse of material goods in a society is called:
    a. exploitation
  * b. recycling
    c. reclamation
    d. restoration
    e. preservation

34. Sustainable development is not a new concept. Which of the following United States presidents alluded to the concept in a speech 85 years ago?
    a. William McKinley
    b. William Taft
    c. Woodrow Wilson
  * d. Theodore Roosevelt
    e. Warren Harding

35. Population stabilization and growth management are necessary because human populations may exceed the _____ of the environment.
    a. infrastructure
    b. infra-infrastruture
    c. growth potential
  * d. carrying capacity
    e. physical infrastructure

36. The Iroquois people speak of the environmental responsibility to provide and care for the seventh generation yet unborn in their everyday lives. The Iroquois are expressing which of the following principles of sustainable development?
    a. ecological justice
    b. recycling
  * c. intergenerational equity
    d. energy efficiency
    e. end-of-pipe controls

37. Soil erosion on a Kansas farm may cause siltation in a downstream reservoir in Oklahoma. This example would be an environmental issue concerning:
    a. ecological justice
    b. intergenerational equity
  * c. intragenerational equity
    d. exceeding the carrying capacity
    e. end-of-pipe pollution control

38. Overfishing in the North Atlantic may deplete some fish stocks for generations to come. This would be an example of an environmental issue concerning:
    a. ecological justice
  * b. intergenerational equity
    c. intragenerational equity
    d. interdependence
    e. participation

39. The use of catalytic converters on automobiles reduces emissions of hydrocarbons and carbon monoxide. Which of the following greenhouse gases is not removed by catalytic converters?
    a. water vapor
  * b. carbon dioxide
    c. nitrogen
    d. oxygen
    e. helium

40. The cutting of old growth forest threatening the habitat of endangered species, such as the spotted owl, would be an example of an environmental issue concerning:
  * a. ecological justice
    b. intergenerational equity
    c. intragenerational equity
    d. carrying capacity
    e. biophysical limits

2
Thinking Critically about Environmental Issues and Solutions

Chapter Outline

Science and the Scientific Method
Hypotheses
Testing Hypotheses
Scientific Theories and Paradigms
Science and Values

Critical Thinking Skills
The Rules of Critical Thinking

Key Terms

hypothesis
inductive reasoning
deductive reasoning
experimental group
epidemiology
paradigm

critical thinking
anecdotal information
correlation
dualistic thinking
scientific theory
paradigm shift

Objectives

1. Discuss the procedure of the scientific method.
2. Define the following terms: "hypothesis", "experiment", "inductive reasoning", "deductive reasoning", "experimental group", "control group" and "independent variable".
3. Define "scientific theory" and discuss how theories are formed.
4. Define "paradigm" and "paradigm shift".
5. Discuss why paradigms are important in the sciences and other disciplines.
6. List and discuss the rules of critical thinking.

Lecture Outline

I. Science and Scientific Method
   A. Science - Science is a body of knowledge and a method of acquiring further knowledge about the world around us.
      1. The term *science* comes from the Latin word *scienta* which means to know or to discern.
      2. Technically, the term *science* refers to a body of knowledge derived from observation, measurement, study, experimentation and the process of accumulating such knowledge.
   B. Scientific Method
      1. Hypotheses - Hypotheses are tentative explanations of observations and measurements.
         a. Inductive Reasoning - Observations and facts are used to arrive at general conclusions.
         b. Deductive Reasoning - A specific conclusion is derived from general principles.
   C. Testing Hypotheses
      1. Hypotheses are tested by performing experiments. Experiments often use two test groups:
         a. Experimental Group - This group is the one that is tested under a new condition; some variable is being altered in this group to test for change.
         b. Control Group - This group is identical to the experimental group but is not exposed to the "new" condition (the variable being tested in the experimental group).
         c. The variable being tested is called the *independent variable.*
   D. Scientific Theories and Paradigms
      1. Theories are explanations that account for many different hypotheses, observations, and facts. Theories are broad scientific generalizations regarding natural, physical and chemical phenomena. A good example would be Atomic theory. Atomic theory explains the structure of the atom and fits numerous observations.
      2. Paradigms are basic models of reality. These are large philosophical and theoretical frameworks which greatly influence the development of theories, laws and generalizations. Paradigms influence our way of thinking and interpreting our observations. A biological example would be the theory of evolution. Paradigms are rarely questioned and become the accepted format for interpreting all new thought or observation.
      3. *Paradigm shifts* occur when new observations cause a change in our fundamental belief system.
   E. Science and Values
      1. Human values shape scientific interpretation and science can influence human values.

II. Critical Thinking Skills
   A. Critical thinking is the ability to distinguish between beliefs and knowledge. Critical thinking helps us separate judgment from facts. These skills are valuable in analyzing a wide range of environmental problems, issues, and information.
   B. The following are basic critical thinking skills:
      1. Gather all information.

2. Understand all terms.
3. Question the methods.
4. Question the source.
5. Question the conclusions.
6. Tolerate uncertainty.
7. Examine the big picture.

Suggestions for Presenting the Chapter

- Performing a simple experimental exercise with the class emphasizing the development and testing of hypotheses is valuable.
- Instructors should provide examples of paradigm shifts to the class for discussion. Some examples might be: Einstein's Theory of Relativity, or the comparison of hunting/gathering societies to modern industrial/post industrial cultures.
- The paradigm shift from a Frontier Ethic to Sustainable Ethics is an interesting discussion and also involves questions of ethics/values.
- The instructor can provide a reading to be analyzed using critical thinking skills. Breaking the class into groups also works well for this type of activity.
- Recommended web sites:

Critical Thinking on the Web: http://www.philosophy.unimelb.edu.au/reason/critical/
The Center for Critical Thinking: http://www.criticalthinking.org/default.html
Scientific Method Quiz:
http://antoine.frostburg.edu/chem/senese/101/intro/scimethod-quiz.shtml
The Scientific Method:
http://www.visionlearning.com/myclassroom/courses/NSC107_Carpi_index.htm
Paradigm Links: http://www.kheper.auz.com/paradigms/links.htm

True/False Questions

1. ___ Science is a collection of belief systems explaining the modern world.
2. ___ Inductive reasoning uses specific fact and observations to arrive at general rules.
3. ___ The control group is the one not exposed to the independent variable.
4. ___ Epidemiology is the study of epidemics.
5. ___ Deductive reasoning provides specific conclusions from general principles.
6. ___ A paradigm is a basic model of reality in science.
7. ___ Theories are scientifically acceptable principles and broad generalizations that account for many facts, observations and hypotheses.
8. ___ Ptolemy hypothesized that the sun was the center of our solar system.
9. ___ Copernicus suggested that the earth was the center of our solar system.
10. ___ Critical thinking is the capacity to distinguish between beliefs and knowledge.
11. ___ Dualistic thinking is encouraged as part of the process of critical thinking.
12. ___ The use of anecdotal information is recommended in good critical thinking decisions.
13. ___ Good critical thinkers examine the "big picture" as a part of the critical thinking process.

14. ___ A correlation is an apparent connection between two variables.
15. ___ Uncertainty is something that must be tolerated in the application of the scientific method.
16. ___ Technology can solve all human problems.
17. ___ All growth is unqualifiably good for humans and the environment.
18. ___ Environmental protection is bad for the economy.
19. ___ The key to success is through the control of nature and natural processes.
20. ___ Critical thinking involves subjecting facts and conclusions to careful analysis before conclusions are made.

True/False Key:

1. F 2. T 3. T 4. T 5. T 6. T 7. T 8. F. 9. F 10. T 11. F 12. F 13. T 14. T 15. T 16. F 17. F 18. F 19. F 20. F

Fill-in-the-Blank Questions

1. _____ comes from the Latin *word scienta* which means "to know or to discern".
2. _____ is a system composed of living things and the interrelated physical and chemical environment.
3. _____ _____ occurs any tie a person uses facts and observations to arrive at general rules or hypotheses.
4. _____ enable scientists to test hypotheses and gain new knowledge.
5. _____ _____ is used when one arrives at a specific conclusion drawn from general principles.
6. Good experiments require control and _____ groups.
7. Scientists design experiments to test the effect of one and only one variable, the_____ variable.
8. _____ is the study of epidemics.
9. Atomic _____ explains the structure of the atom and fits numerous observations.
10. A _____ is a basic model of reality in science.
11. The belief that the earth is the center of the solar system is called the _____ view.
12. _____ _____ is the capacity to distinguish between beliefs and knowledge.
13. A _____ is an apparent connection between two variables.
14. _____ thinking is reasoning that is black-and-white, right-or-wrong-oriented.
15. The sixth rule of critical thinking is to tolerate _____.
16. One myth of modern society is that all _____ is unqualifiably good.
17. One principle of critical thinking is to understand all the _____ and concepts related to the problem.
18. In contract to hypotheses, _____ generally cannot be tested by single experiments.
19. Ptolemy was a _____ astronomer that hypothesized that the earth was the center of our solar system.
20. Good critical thinking requires that one gathers all the _____ you can before you can make a decision.

Fill-in Answers:

1. Science
2. Ecosystem
3. Inductive reasoning
4. Experiments
5. Deductive reasoning
6. experimental
7. independent
8. Epidemiology
9. theory
10. paradigm
11. geocentric
12. critical thinking
13. correlation
14. Dualistic
15. uncertainty
16. growth
17. terms
18. theories
19. Greek
20. information

Multiple Choice Questions

1. A body of knowledge derived from observations , measurement, study, and experimentation and the process of accumulating such knowledge is best called:
   a. theory
   b. observation
   * c. science
   d. experimentation
   e. hypothesis

2. Ecology is the study of:
   a. forests
   * b. ecosystems
   c. plants
   d. animals
   e. humans

3. When specific facts and observations are used to arrive at general rules or hypotheses, _____ reasoning is being used.
   a. deductive
   b. deceptive
   c. inverse
   * d. inductive
   e. circumstantial

4. An explanation that can account for a variety of different observations is called a:
   a. theory
   b. paradigm
   c. fact
   * d. hypothesis
   e. guess

5. A system composed of living things and the interrelated physical and chemical environment is called a (an):
    a. lithosphere
    b. biosphere
    *   c. ecosystem
    d. ecotone
    e. cline

6. Hypotheses are tested by performing:
    *   a. experiments
    b. theories
    c. inductive reasoning
    d. deductive reasoning
    e. paradigm shifts

7. In a study of the effects of air pollution on laboratory mice, the _____ would be exposed to certain amounts of pollution.
    a. control group
    *   b. experimental group
    c. out group
    d. test group
    e. independent group

8. The group of mice not exposed to any pollution in the experiment mentioned in question #7 are the:
    *   a. control group
    b. experimental group
    c. out group
    d. test group
    e. independent group

9. The pollution used in the experiment mentioned in question #7 would be the:
    a. control
    b. dependent variable
    *   c. independent variable
    d. constant
    e. paradigm

10. When general principles are used to arrive at a specific conclusion _____ reasoning is being used.
    *   a. deductive
    b. deceptive
    c. inverse
    d. inductive
    e. circumstantial

11. Philosophical and theoretical frameworks within which theories, laws, and generalizations are formulated are called:
    a. theories
    b. hypotheses
    *   c. paradigms
    d. studies
    e. value systems

12. The capacity to distinguish between beliefs and knowledge is a skill called:
    a. paradigm shift
    b. values clarification
    c. analysis
    d. synthesis
    *   e. critical thinking

13. Which of the following responses lists the normal order of the scientific method?
    a. hypothesis, observation, experiment
    b. hypothesis, experiment, observation
    c. observation, experiment, hypothesis
    *   d. observation, hypothesis, experiment
    e. experiment, observation, hypothesis

14. Experiments are formulated to test which of the following components of the scientific method:
    a. observation and measurement
    *   b. hypothesis
    c. experiment
    d. theory formation
    e. substantiated hypothesis

15. Ptolemy, an ancient Greek astronomer, hypothesized that the _____ was the center of the solar system.
    a. sun
    b. moon
    *   c. earth
    d. Mars
    e. Jupiter

16. Copernicus hypothesized that the _____ was the center of the solar system.
    *   a. sun   d. Mars
    b. moon   e. Jupiter
    c. earth

17. A scientist observes three brightly colored species of snakes all of which are poisonous. The scientist may suggest that all brightly colored snakes are poisonous by using _____.
    a. deductive reasoning
* b. inductive reasoning.
    c. experimentation.
    d. observation and measurement
    e. critical thinking

18. The stage of the scientific method characterized by the testing of hypotheses is:
    a. observation and measurement
    b. hypothesis formation
* c. experimentation
    d. revision of hypotheses
    e. formation of general theories

19. If a hypothesis is not verified by an experiment the next step in the scientific method would be:
* a. formation of a new hypothesis.
    b. experimentation
    c. formation of new theories
    d. new observations and measurement
    e. a paradigm shift

20. A group of scientists suggest that rattlesnakes shake their rattles as a warning behavior. The previous statement would be best categorized as a:
    a. observation
* b. hypothesis
    c. theory
    d. experiment
    e. paradigm

21. Brightly colored animals are often poisonous. If you encounter a brightly colored snake you might suggest it is also poisonous using
    _____.
* a. deductive reasoning
    b. inductive reasoning
    c experimentation
    d. observation and measurement
    e. critical thinking

22. The *heliocentric view* of the solar system was first suggested by:
    a. Copernicus
    b. Ptolemy
    c. Newton
    d. Einstein
* e. early Greek astronomers

23. Which of the following characteristics would **not** be a critical thinking skill:
    a. gathering all information
    b. expect and tolerate uncertainty
    c. examine the big picture
    d. question the source of the information
* e. accept all the information/facts

24. Which of the following is **not** a myth about modern society?
    a. Individual actions do not count.
    b. Technology can solve all our problems.
    c. People are separate from nature.
* d. We have obligations to future generations.
    e. Happiness stems from material possession.

25. The *geocentric view* of the solar system was first suggested by:
    a. Newton
    b. Aristotle
    c. Copernicus
* d. Ptolemy
    e. Pliny

3

Understanding the Root Causes of the Environmental Crisis and Solutions

Chapter Outline

The Roots of the Environmental Crisis
> Religious Roots
> Cultural Roots: Democracy and Industrialization
> Biological and Evolutionary Roots
> Psychological and Economic Roots

Leverage Points
> Making Human Systems and Technologies Sustainable
> Changing Our Perceptions, Values and Beliefs
> Unsustainable Ethics

Key Terms

root causes
Industrial Revolution
frontierism
biological imperialism
frontal lobe
ethics
frontier ethic
ecological backlashes
sustainable ethics

Objectives

1. Discuss the religious roots of the environmental crisis.
2. Discuss the cultural roots of the environmental crisis.
3. Discuss the psychological and economic roots of the environmental crisis.
4. Discuss the biological roots of the environmental crisis.

5. Review the actions that could remedy the root causes of the modern environmental crisis.
6. Summarize the difference between "frontier ethics" and "sustainable ethics".

Lecture Outline

I.   Roots of the Environmental Crisis
 A. Religious Roots
  1. Lynn White's thesis was that the roots of the environmental crisis are largely religious and stem from the early influence on western culture of Christianity. He argued that western-style science and technology "got their start, acquired their character, and achieved world dominance in the Middle Ages." The most important influence on science and technology at this time was Christianity's view of human dominance over nature. White argued that during the Middle Aged Christianity replaced paganism. Pagans viewed themselves as part of the whole and Christianity promoted a sense of dualism, a view of humanity as separate from nature.
  2. Lewis Moncrief pointed out that religious belief certainly influences human behavior towards the environment but to argue that religion is the primary cause of environmental destruction is not supported by the facts. Other civilizations have altered their environments severely long before the influence of Christianity. Religion is likely a powerful influence of belief systems but not the sole root cause of environmentally destructive behavior.
 B. Cultural Roots: Democracy and Industrialization
  1. Democracy - The influence of democratic ideals after the French Revolution had a significant impact on how the environment was used. Prior to the French Revolution most land and resources in the Western world were owned primarily by royalty or the Roman Catholic Church. The French Revolution marked the demise of feudalism and this system of land ownership. Land became vested in the hands of the many.
  2. Industrialization - The shift from manual labor to energy-intensive machine production is an important force affecting the environment. Moncrief argues that the success of democracy and industrialization resulted in a more equitable distribution of wealth among the human population and , more important, rising affluence and consumption. Pollution and environmental destruction were the results.
  3. Frontierism - Movement of people to farm and homestead undisturbed land fostered the philosophy of *frontierism*. *Frontierism* views the environment as an unlimited resource. Wilderness was an obstacle to overcome during the process of settlement. Clearing a forest or draining a marsh is acceptable behavior because of the perception that these resources are unlimited and therefore of little immediate concern or value.
 C. Biological and Evolutionary Roots
  1. Biological Imperialism - The natural tendency of populations to expand in proportion to the resources available.
   a. Major environmental transgressions of the past and the present may originate in human biological imperialism.
   b. The development of technologies have permitted us to expand almost uncontrollably.
 D. Psychological and Economic Roots

1. Human attitudes and beliefs are responsible for many unsustainable practices.
   a. Denial - Denial of the reality of our environmental dilemma may delay appropriate action to correct our behavior.
   b. Apathy
   c. Greed
   d. Acquisitiveness
   e. Inability to respond to subtle threats. Our brain has evolved to respond to immediate, life threatening situations.
2. Economic beliefs have fostered unsustainable behavior and have not tempered our biological imperialistic tendencies.
   a. The unquestioned belief that economic growth is always good.
   b. People consume to make themselves feel better. Advertisers promise that the goods we purchase will make us happy or successful.
   c. People are encouraged to consume to conform to the norms of society. In order to belong, one must have the right car, clothes, the right cologne or perfume, etc.
   d. Consumption also is encouraged by the joy of novelty and the usefulness and convenience of many new products.

II. Leverage Points
   A. Making Human Systems and Technologies Sustainable
      1. Human technologies and human systems are fundamentally unsustainable.
   B. A New Worldview: Changing Our Perceptions, Values, and Beliefs
      1. Unsustainable Ethics - In most western nations, human values express a kind of cavalier frontierism that hampers efforts to create a sustainable future.
      2. The First Tenet of Unsustainable Ethics - The Earth is an unlimited supply of resources for exclusive human use.
      3. The Second Tenet of Unsustainable Ethics - Humans are not a part of nature and are immune to natural laws.
      4. The Third Tenet of Unsustainable Ethics - Human success derives from efforts to subdue and control nature.
      5. Some Impacts of Frontier Thinking
         a. Frontier ethics profoundly affects how people act.
         b. Frontier ethics influences how people solve environmental problems.
         c. Frontier ethics influences our personal goals and expectations and drives rampant consumerism and unnecessary use of resources.
      6. Sustainable Ethics
         a. The Earth has a limited supply of resources and they are not all for us.
         b. Humans are a part of nature and subject to its laws.
         c. Success stems from efforts to cooperate with the forces of nature
         d. Our future depends upon creating and maintaining a healthy, well-functioning global ecosystem.

Suggestions for Presenting the Chapter

- Exercises in value identification and clarification are useful at this stage in the course. Topics might include an examination of the importance of "wilderness" or "recycling" in their lives. This could be contrasted with the desire to acquire a new car or large home or even a new pair of designer jeans. The idea is to produce active examination of our values, goals and aspirations and how these may not consider issues of sustainability.

- Students with experience in American History may be asked about the concept of "manifest destiny". This leads very naturally into a discussion of the "frontier ethic". Introduction of aboriginal/Native American readings as a contrast in values from the same time period allows contrast and comparison of the frontier mentality to a sustainable, earth-based ethical system.

- A review of the natural history of your area since the formation of the United States is an interesting exercise. The loss of natural diversity in many areas illustrates to students the scope of our environmental impacts. Many students do not realize the immense ecological changes that have occurred as a result of development. A historical perspective often demonstrates effectively the impact of "taming the frontier" and its legacy.

- Instructors should emphasize that everyone has an impact on the environment. Often the most positive impact they can have is on their friends. Students should be encouraged to interact with their family and friends outside of class time about issues of environmental ethics. This interaction plays an important role in encouraging critical thinking and value clarification for the student, family members, and friends.

- Recommended web sites:

Brief History of the Origins of Environmental Ethics: http://www.cep.unt.edu/novice.html
Center for Applied Ethics: http://www.ethics.ubc.ca/resources/environmental/
The Historical Roots of Our Ecological Crisis: http://www.zbi.ee/~kalevi/lynwhite.htm
Earth Ethics Quarterly: http://www.center1.com/ethics.html
Journal of Environmental Ethics: http://www.cep.unt.edu/enethics.html

True/False Questions

1. ___ The current environmental crisis is a crisis of unsustainability.
2. ___ Lewis Moncrief attributes the environmental crisis to the spread of democracy and industrialism.
3. ___ The shift from manual labor to the use of machines is called the industrial revolution.
4. ___ Frontierism assumes that there is an inexhaustible reservoir of nature resources to be exploited.
5. ___ The innate ability of living populations to expand to fill their environments is called biological imperialism.
6. ___ Human technology has allowed our species to expand almost uncontrollably.
7. ___ Humans are very poor colonizers of most environments on earth.
8. ___ Denial is a psychological factor that contributes to our inability to respond to the environmental crisis.
9. ___ The "frontier ethic" suggests that the earth is an unlimited supply of resources for exclusive human use.
10. ___ Sustainable ethics suggests that we must fight nature and natural systems to survive.
11. ___ An ecosystem consists of a community of organisms and all of the interactions between them and their environment.

12.___ Ecological backlashes occur when we use the principles of sustainability in dealing with natural systems.

13.___ The first tenet of sustainable ethics is that the earth is an unlimited source of resources for human use.

14.___ Paul Wachtel suggested that the modern culture of mass consumption is developed around a core of unfulfilled longing.

15.___ Andrew Schmookler observed that growth is "an omnipresent symbol of the good".

16 ___ The tragedy of the commons occurs when community resources are abused by individuals interested in short-term profits.

17.___ Human technologies and human systems are fundamentally unsustainable.

18.___ Ethics refers to the system of values that determine personal action and behavior.

19.___ Humans are a part of nature and subject to its laws.

20.___ Pesticides are used to increase insect populations on farms.

True/False Key:

1. T 2. T 3. T 4. T 5. T 6. T 7. F 8. T 9. T 10. F 11. T 12. F 13. F 14. F 15. F 16. T 17. T 18. T 19. T 20. F

Fill-in-the-Blank Questions

1. Lynn White argued that the roots of our environmental crisis are largely _____.

2. Moncrief argued that the environmental crisis is due to the spread of _____ and _____.

3. The shift from manual labor to energy-intensive machine is called the _____ _____.

4. Biological _____ is the tendency for all living organisms to expand their populations to   fill the environment they live in.

5. The _____ _____ was introduced in 1884 from South American and has invaded many waterways in the southern United States.

6. The world's environmental woes result from human technologies and human systems that are fundamentally _____.

7. An ecosystem consists of a community of _____ and all of the interactions between them and their physical environment.

8. _____ are chemicals used to kill organisms that farmers view as pests.

9. Pest resistance that has developed as a result of pesticide application is a good example of an ecological _____.

10. Lynn White noted that during the Middle Ages, Christianity promoted a sense of _____, a view that humanity is separate from nature.

11. _____ is the idea the world is an unlimited supply of resources created solely for human use.

12. The _____ _____ is the part of the forebrain that allows humans to plan and think about the future.

13. _____ have permitted the human population to expand almost uncontrollably.

14. As humans spread into the North American continent they may have been responsible for the _____ of the megafauna.
15. _____ is an important psychological factor that contributes to our inability to respond to the environmental crisis.
16. People came across the Bering Strait land bridge to North American about _____ years ago.
17. Paul Ehrlich and Robert Ornstein suggest that the human nervous system is built to respond to _____ physical danger.
18. Our early evolutionary success is due in part to social groups that permitted cooperative _____ and food-gathering ventures and joint efforts to ward off predators.
19. The abuse of publicly owned resources by individuals for short-term gain is best called the tragedy of the _____.
20. In ancient Greece, _____ recognized that property shared freely by many people often received the least care.

Fill-in Key:
1. religions
2. democracy, industrialization
3. industrial revolution
4. imperialism
5. water hyacinth
6. unsustainable
7. organisms
8. Pesticides
9. backlash
10. dualism
11. frontierism
12. frontal lobe
13. technologies
14. extinction
15. denial
16. 10,000
17. immediate
18. hunting
19. commons
20. Aristotle

Multiple Choice Questions

1. Lynn White argued that the most important influence of science and technology in the Middle Ages was _____ view of human dominance over nature.
   a. feudalism's
   b. Buddhism's
   * c. Christianity's
   d. Aristotle's
   e. Plato's

2. The shift from manual labor to machine production is called the:
   a. New Paradigm
   * b. Industrial Revolution
   c. Agricultural Revolution
   d. Reformation
   e. Renaissance

3. The ability of populations to expand in proportion to their resources is:
   a. expansionism
   * b. biological imperialism
   c. Darwinism
   d. exponential growth
   e. commensalism

4. The early inhabitants of North America migrated from Asia across the Bering land bridge about _____ years ago.
   a. 500
   b. 1000
   c. 5000
   * d. 10000
   e. 50000

5. Technological development was made possible by two evolutionary developments:
   a. manual dexterity and bipedal locomotion
   b. bipedal locomotion and cranial size
   c. cranial size and arm length
   * d. manual dexterity and brain development
   e. bipedal locomotion and reproductive rate

6. Prior to the French Revolution, much of the land and resources in Europe were owned primarily by royalty or by the:
   a. government
   b. people
   * c. church
   d. cities
   e. wealthy

7. The idea that resources are inexhaustible and are an obstacle to overcome is best described by which of the following responses?
   a. biological imperialism
   b. imperialism
   * c. frontierism
   d. environmentalism
   e. materialism

8. Which of the following statements is **not** part of the "frontier ethic"?
   a. The earth has unlimited resources.
   b. Resources are exclusively for human use.
   c. Humans are not part of nature and are immune to her laws.
   d. Human prosperity and well-being result from our control of nature.
   * e. We share the earth with other creatures.

9. Phytoplankton in ancient seas is found in chemicals of which of the following modern products?
   a. soap
   * b. plastics
   c. wool
   d. hemp
   e. bubble gum

10. Which of the following would not be a factor that could contribute to the modern environmental crisis?
    a. denial
    b. apathy
    c. inability to respond to subtle threats
    d. greed
 * e. frugality

11. The concept that property shared freely by many often received the least care is called the:
    a. leverage point
    b. survival of the fittest
 * c. tragedy of the commons
    d. common good
    e. ecological backlash

12. Who said, "Nature and man can never be friends. Fool, if thou canst not pass her, rest her slave!?"
    a. Paul Ehrlich   *d. Matthew Arnold
    b. Paul Wachtel   e. Thomas Kean
    c. Garrett Hardin

13. Many environmental problems today result from our efforts to control and dominate nature. These impacts are called:
    a. biological imperialism
 * b. ecological backlashes
    c. tragedy of the commons
    d. leverage
    e. root causes

14. The personal beliefs which define right and wrong behavior are called:
 * a. ethics
    b. laws
    c. rules
    d. instincts
    e. adaptations

15. This UCLA historian wrote a paper entitled "The Historical Roots of Our Ecological Crisis."
    a. Paul Ehrlich
 * b. Lynn White
    c. Gabriel Fackre
    d. Lewis Moncrief
    e. Bertrand Russell

16. Which of the following responses is a characteristics of sustainable ethics?
    a. The earth has a limited supply of resources and they are not all ours.
    b. Humans are a part of nature.
    c. Success stems from efforts to cooperate with nature.
    d. Our future depends on creating and maintaining healthy ecosystems.
 * e. all of the above

17. A community of organisms and all of the interactions between them and their physical environment is a/an:
    a. biome
    b. habitat
 * c. ecosystem
    d. biosphere
    e. ecotone

18. Which of the following is **not** a root cause of the environmental crisis?
    a. spread of democracy
    b. industrialization
    c. biological imperialism
    d. religious beliefs
 * e. sustainable ethics

19. Andrew Bard Schmookler, contends the modern culture of mass consumption is developed around a core of:
    a. greed and corruption
    b. competitive growth
    c. narcissistic egotism
 * d. unfulfilled longing
    e. romantic fronticrism

20. This author wrote *The Poverty of Affluence* and said it is "very difficult for us to accept any idea of a limit to growth as implying anything other than stagnation."
    a. Paul Ehrlich
    b. Andrew Bard Schmookler
    c. Lewis Moncrief
    d. Gabriel Fackre
 * e. Paul Wachtel

21. This philosopher wrote that "every living thing is a sort of imperialist, seeking to transform as much as possible of the environment into itself and its seed."
    a. Paul Ehrlich
    b. Lynn White
    c. Gabriel Fackre
    d. Lewis Moncrief
  * e. Bertrand Russell

22. This plant was native to South America and since its introduction into the United States in 1884 has clogged many important waterways.
    a. kudzu
    b. pondweed
    c. wild rice
  * d. water hyacinth
    e. cattail

23. A huge portion of the oxygen we breathe each year is replenished by:
    a. decomposition
    b. anaerobic fermentation
  * c. plants and algae
    d. zooplankton
    e. marine evaporation

24. The loss of tropical rain forests has long-term impacts such as:
  * a. climate change
    b. ozone depletion
    c. loss of grazing land
    d. loss of grasslands
    e. sinkhole formation

25. The statement "A human apart from environment is an abstraction - in reality no such thing exists", was made by:
    a. Albert Camus
  * b. Raymond Dasmann
    c. Paul Ehrlich
    d. Gabriel Fackre
    e. Lynn White

4

Environmental Protection by Design: Restructuring Human Systems for Sustainability

Chapter Outline

Human Settlements: Networks of Systems

Why Are Human Systems Unsustainable?

Making Systems Sustainable: Applying the Principles of Sustainable Development
What are the guiding principles to sustainable development?
Applying the Principles of Sustainable Development to Human Systems

Key Terms

energy supply system
food production system
waste management system
manufacturing system
transportation system
economic system
system of government
educational system

biological system
sustainable development
primary input
secondary throughput
primary output
operating principles of sustainable development

Objectives

1. Summarize the systems found in a human settlement.

2. Discuss why human systems are often unsustainable.
3. Discuss the operating principles of sustainable development.
4. Discuss how the principles of sustainable development can be applied to your home.
5. Summarize what you can do to apply the principles of sustainability to your lifestyle.

Lecture Outline

I. Human Settlements: Networks of Unsustainable Systems
   A. Human settlements composed of many interdependent systems such as transportation, energy, and waste management. Growing evidence shows that these systems are not sustainable in the long run.

II. What's Wrong with Human Systems: Why Are They Unsustainable?
   A. Although human systems may provide us with a steady stream of goods and services, they are systematically reducing the carrying capacity of the planet. Human systems:
      1. Produce pollution in excess of the planet's ability to absorb and detoxify wastes.
      2. Deplete nonrenewable resources faster than substitutes can be found.
      3. Use renewable resources faster than they can be regenerated.

III. Making Systems Sustainable: Applying the Principles of Sustainable Development
   A. The following are directive or guiding principles of sustainable development:
      1. Population stabilization.
      2. Growth management.
      3. Efficiency.
      4. Renewable energy use.
      5. Recycling.
      6. Restoration.
      7. Sustainable management.

IV. Applying the Operating Principles to Human Systems
   A. Human systems can be revamped by using the operating principles for redesign and restructuring.
   B. The primary community systems are:
      1. Transportation (surface and water-based systems).
      2. Housing and other buildings (construction systems)
      3. Agriculture, food processing and distribution (food production system)
      4. Business (commercial and industrial)
      5. Educational
      6. Governmental
   C. The systems above can be improved/revamped by applying the following principles of sustainable development to each system and devising an action plan (this would be done on the local, state or national level):
      1. Conservation (efficiency and frugality).
      2. Recycling and composting.
      3. Renewable resource use (especially energy).
      4. Habitat protection, restoration, and sustainable management.
      5. Growth management (controlling growth patterns and stabilizing growth).

Suggestions for Presenting the Chapter

- Before presenting the chapter have students break into groups and list the human systems in the community they live in. This provides an interesting comparison to what you will do in lecture and also provides allows a student directed activity to start the class with. This fosters a participatory attitude necessary for community activism.
- After delivering the chapter lecture break the class into work groups and have them work on the chart provided in Figure 4-3 of the text. Educational and governmental systems can be added to the list of systems in the left hand column.
- Have your students collect articles from the newspaper that relate to the issue of human systems and sustainability. They should focus on changes at the local, state or national level in systems that are unsustainable or foster sustainability.
- Have your students analyze the educational system. Does your institution have an environmental policy? What is done on campus to promote sustainability? Does your campus have any student environmental organizations or an environmental club?
- Recommended web sites:

Center of Excellence for Sustainable Development: http://www.sustainable.doe.gov/
United Nations Sustainable Development: http://www.un.org/esa/sustdev/
Virtual Library/Sustainable Development: http://www.ulb.ac.be/ceese/meta/sustvl.html
World Resources Institute: http://www.wri.org/wri/materials/index.html
President's Council on Sustainable Development:
        http://www.whitehouse.gov/PCSD/Publications/suscomm/ind_suscom.html

True/False Questions

1. ___ The first major societal change in human populations was the Industrial Revolution.
2. ___ The agricultural revolution started about 10,000 B.C..
3. ___ The industrial revolution started in England in the late 1800s.
4. ___ The industrial revolution was driven by the use of fossil fuels like coal.
5. ___ The transmission lines, power plants and coal mines are part of the energy supply system.
6. ___ The farms, food processing facilities and grocery stores are part of the manufacturing system.
7. ___ Human systems are supported by the infra-infrastructure of society, the biological system.
8. ___ Global warming is due to the accumulation of gases like carbon dioxide in the atmosphere.
9. ___ Human systems are unsustainable because they produce pollution in excess of the planet's ability to absorb and detoxify waste.
10. ___ Humans do not need to stabilize their population growth to act in a sustainable fashion.
11. ___ Human energy systems should be restructured to use solar and wind energy and abandon fossil fuels to become sustainable.
12. ___ Gray water is water from toilets and outhouses.

13 ___ Photovoltaic cells generate electricity when light strikes the cells.
14. ___ Sustainable housing can be built largely from recycled materials like tires and insulation.
15. ___ Recycling is an important principle of sustainable development.
16. ___ Installing water-efficient shower heads will make your water usage more sustainable.
17. ___ Conservation is an important principle of sustainable development.
18. ___ Business and industry should be exempted from attempts at sustainability because they already are sustainable endeavors.
19. ___ Coal is one of the most important renewable resources.
20. ___ Human systems exceed the carrying capacity of the planet to supply resources and deal with our wastes.

True/False Key:

1. F 2. T 3. F 4. T 5. T 6. F 7.T 8. T 9. T 10. F 11. T 12. F 13. T 14. T 15. T 16. T 17. T 18.F 19. F 20. T

Fill-in-the-Blank Questions

1. The shift from hunting and gathering to subsistence farming is called the _____ revolution.
2. The agricultural revolution started about _____ years ago.
3. Oil and natural gas are _____ fuels.
4. The industrial revolution started in England in the _____ s.
5. The infrastructure and system for transporting people and goods is the _____ system.
6. The _____ system provides the resources for recycling our wastes.
7. Farms, food processing facilities and grocery stores are part of an elaborate _____ production system.
8. Human settlements are made up of many _____ systems.
9. The warming of the earth's atmosphere due to the accumulation of gaseous pollutants is called _____ warming.
10. Human systems are unsustainable because they deplete finite _____ resources faster than they can naturally regenerate.
11. One of the principles of sustainable development is to _____ natural systems that have been damaged in years past.
12. Part of recycling is the purchase of _____ products.
13. Sustainable population stabilization means that we should support _____ planning.
14. Wind generation of power would be an example of a _____ resource.
15. Installing compact fluorescent light bulbs is part of _____ conservation.
16. Air pollution and loss of open space are _____ of unsustainable systems.
17. Water from the shower and washing machine is called _____ water.
18. It is suggested that the temperature setting during the summer on thermostats should be set at ____ degrees Fahrenheit.

19. Another name for solar cells is _____ .
20. In windy areas, a small wind _____ could be used to supplement your home energy needs.

Fill-in Answers:

1. agricultural
2. 10,000
3. fossil
4. 1700
5. transportation
6. biological
7. food
8. interdependent
9. global
10. renewable
11. restore
12. recycled
13. family
14. renewable
15. energy
16. symptoms
17. gray
18. 78
19. photovoltaic
20. generator

Multiple Choice Questions

1. The long, progressive shift from hunting and gathering to subsistence-level farming that started about 10,000 years ago is the:
   * a. Agricultural Revolution
   b. Industrial Revolution
   c. Sustainable Revolution
   d. Food Production System
   e. Fertile Crescent

2. The power plant, transmission lines, coal mines and house wiring are part of the:
   a. biological system
   b. economic system
   c. manufacturing system
   * d. energy system
   e. waste management system

3. Which of the following is **not** a principle of sustainable development?
   a. Population stabilization
   b. Management of growth
   c. Efficiency
   d. Renewable energy use
   * e. Landfilling reusables

4. Which of the following is **not** a personal action for creating a more sustainable home?
   a. installing insulation
   b. caulk and weather-strip
   c. install ceiling fans
   * d. remove insulation from hot water heater
   e. buy energy-efficient appliances

5. Which of the following are ways to use water more sustainably at home?
   a. Install water-efficient shower heads
   b. Take shorter showers
   c. Take showers rather than baths
   d. Don't shave with the water running
   * e. all of the above

6. Which of the following responses is a reason human systems are unsustainable?
   a. They use renewable resources slower than they can regenerate.
   b. They do not deplete nonrenewable resources.
   * c. They produce levels of pollution that exceed the global capacity to absorb and render them harmless.
   d. Resources are recycled in closed-loop systems.
   e. Restoration projects are actively pursued.

7. Renewable energy supplies would include which of the following energy sources?
   a. coal
   b. natural gas
   c. oil
   * d. solar
   e. coal tar

8. This system provides the goods and services and serves as a sink for our wastes:
   a. manufacturing system
   b. health care system
   * c. transportation system
   d. biological system
   e. economic system

9. All human settlements are made up of many _____ systems.
   a. independent
   * b. interdependent
   c. sustainable
   d. information
   e. output

10. To create a sustainable future, the challenge is twofold: to build new systems using the principles of sustainable design and to:
    a. eliminate current systems
    b. continue operating unsustainable systems
    * c. retrofit existing systems with sustainable alternatives
    d. decrease efficiency of all systems
    e. reduce the amount of recycling

11. This human system consists of dumps, recycling facilities, incinerators, compost facilities and trucks that move materials from place to place.
    a. economic system
    b. information system
    c. transportation system
    * d. waste management system
    e. biological system

12. The release of this substance into the atmosphere is causing the earth's temperature to increase:
    a. ozone
    b. oxygen
    c. nitrogen
    * d. carbon dioxide
    e. radon

13. This human system provides training for people to function in other human systems:
    a. transportation system
    b. system of government
    * c. educational system
    d. manufacturing system
    e. waste management system

14. This human system produces, meat, grains dairy products, and sells them to consumers:
    a. waste management system
    * b. food production system
    c. manufacturing system
    d. transportation system
    e. information system

15. The current transportation system is unsustainable primarily because of its use of:
    a. solar energy
    b. renewable energy sources
    * c. fossil fuels
    d. nuclear energy
    e. fuel cells

16. Which of the following responses is not a personal action for promoting sustainable restoration?
    a. Return part of your lawn to wildlife habitat.
    b. Plant native vegetation to reduce water use.
    c. Replant areas that have not regrown.
    d. Buy wood from sustainable harvested forests.
    * e. Install an energy-efficient wood stove.

17. Primary output from human systems will have to be removed and /or recycled by the _____system.
    a. transportation
    b. energy
    * c. waste management
    d. government
    e. housing

18. Human systems may provide us with a steady stream of goods and services but they are systematically reducing the _____ of the planet.
    a. renewable resources
    b. energy efficiency
    c. reproductive potential
    * d. carrying capacity
    e. environmental damage

19. Which of the following activities would **not** be a way to achieve a sustainable transportation system?
    * a. Change to diesel powered autos.
    b. Use fuel-efficient cars.
    c. Use more solar powered cars.
    d. Support carpooling
    e. Support efforts to expand mass transit.

20. People and goods are moved via an extensive network of flight paths, highways, railways, and navigable waterways. This is the human:
    a. information system
    * b. transportation system
    c. manufacturing system
    d. waste management system
    e. energy system

21. Which of the following is **not** a principle of sustainable development?
    a. Sustainable management
    b. Restoration
    c. Recycling
    d. Renewable resource use
  * e. Fossil fuel use

22. Adding a wind generator to your house would enhance your sustainable use of:
    a. water
    b. recycling
    c. population stabilization
  * d. renewable resources
    e. restoration projects

23. Limiting your family size and supporting family planning efforts is sustainable activity aimed at reducing:
    a. energy use
    b. water
    c. population growth
    d. resource depletion
  * e. all of the above

24. Living close to work and carpooling are sustainable solutions to reducing:
    a. population size
    b. water use
  * c. energy use
    d. renewable resources
    e. restoration efforts

25. Which of the following items would you see in a sustainable home?
    a. passive solar design
    b. energy-efficient appliances
    c. superinsulated ceilings and walls
    d. gray water system
  * e. all of the above

26. Air pollution, water pollution, and loss of natural habitats are all symptoms of:
    a. sustainable development
  * b. unsustainable systems
    c. normal human activity
    d. fossil fuel use
    e. population decline

5

Principles of Ecology: How Ecosystems Work

Chapter Outline

Humans and Nature: The Vital Connections
    Are humans a part of nature?

Ecology: The Study of Natural Systems
    What is ecology?

The Structure of Natural Systems
    The Biosphere and the Importance of Recycling and Renewable Resources
    Biomes and Aquatic Life Zones
    Ecosystems: Abiotic and Biotic Components

Ecosystem Function
    Food Chains and Food Webs
    The Flow of Energy and Matter Through Ecosystems
    Trophic Levels
    Nutrient Cycles
    The Carbon Cycle
    The Nitrogen Cycle

Key Terms

ecology	tundra	detritivores
abiotic	taiga	food web
biotic	temperate deciduous forest	biomass
biosphere	grassland	nutrient cycle
photosynthesis	range of tolerance	nitrogen fixation
closed system	zones of physiological stress	competitive exclusion principle
biome	zones of intolerance	biomass pyramid
aquatic life zone	endangered species	energy pyramid
limiting factor	population	pyramid of numbers
niche	competition	carnivores
consumers	producers	herbivores
omnivores	food chain	

Objectives

1. Define "biosphere", "biome", and "ecosystem".
2. Discuss "abiotic and biotic factors" and their significance in ecosystems.
3. Define the terms "range of tolerance", "zones of physiological stress" and "zones of intolerance".
4. Compare the terms "habitat" and "niche".
5. Discuss the competitive exclusion principle.
6. Diagram and label the trophic levels of a typical food web.
7. Explain why there is a pyramid of biomass, energy and numbers in ecosystems.
8. Discuss the carbon and nitrogen cycles.
9. List the major biomes and, for each, give distinguishing characteristics and identify the major threats.
10. Explain why humans are a part of nature and are dependent on natural systems for survival.

Lecture Outline

I. Humans and Nature: The Vital Connections
   A. Humans are a part of nature. Like all other species, humans depend on the soil, air, water, sun and a host of living organisms to survive.

II. Ecology: The Study of Natural Systems
   A. Ecology is the study of organisms and their relationships to one another and to the environment.
   B. Environment is a term referring to our surroundings; it is not synonymous with ecology.

III. The Structure of Natural Systems
   A. The biosphere or ecosphere refers to the life-supporting portion of the earth. The biosphere is a closed system because, within it, all materials are recycled and reused. Humans break these cycles at great risk.
   B. Biomes are terrestrial areas with distinctive climate, soil characteristics, and plant and animal associations.
      1. Precipitation and temperature are the primary controlling factors in biomes.
      2. North America's five major biomes are: tundra, taiga, temperate deciduous forest, grassland and desert.
   C. Aquatic Life Zones
      1. These zones, similar to biomes, are determined primarily by sunlight penetration and nutrient availability.
      2. Coral reefs, estuaries and coastal wetlands, deep ocean areas, and continental shelves are distinct aquatic life zones.
   D. Ecosystems or ecological systems, are dynamic networks of interdependent plants and animals in a particular environment. Ecotones are transitional areas of overlap between adjacent ecosystems. All ecosystem components are either biotic (living) or abiotic (nonliving).
      1. Abiotic Factors · These include sunlight physical factors, and chemical components. Each organism can survive only within the limits of its range of tolerance to abiotic factor fluctuation and will thrive only within the optimum

range. The limiting factor may be a abiotic element, such as rainfall, temperature, or nitrate, that is primarily responsible for restricting the growth or reproduction of key organisms in an ecosystem.

    2. Biotic Factors · These include all living things in an ecosystem. Groups of the same species in an area form a population; several populations living together form a community. Organisms within a community may interact through predation, commensalism, mutualism, neutralism, parasitism, or inter-/intraspecific competition. Biotic factors can be potent forces in shaping the structure of biological communities.

        a. Humans are one of the most fierce competitors in the biotic community. Our activities effect both the abiotic and biotic components of ecosystems.

IV. Ecosystem Function

  A. Food Chains - A food chain is a series of organisms, each feeding on the preceding. Food chains may be either the grazer or decomposer (detritus) type.

    1. Each organism is either a producer (or autotroph, self-feeder) or a consumer (or heterotroph, other-feeder).

    2. Depending on their primary feeding habits, consumers may be classified as herbivores, carnivores, or omnivores.

  B. Food Webs - A food web is a complex network of feeding interactions involving many organisms and food chains. Food web is a term which more accurately describes the feeding relationships in an ecosystem. Generally, the more complex a food web, the more stable it is.

  C. Classifying Consumers Organisms occupy feeding or trophic levels.

    1. The first trophic level is occupied by producers; the second, third, and subsequent levels are occupied by primary, secondary, etc. consumers.

  D. The Flow of Energy and Nutrients Through Food Webs

    1. All ecosystems ultimately depend on photosynthesis by green plants for their energy. Food chains are biological avenues for the flow of energy and the cycling of nutrients in the environment.

    2. Energy flows in one direction through food chains, but nutrients are recycled.

    3. Nutrients in the food chains reenter the environment through waste or the decomposition of dead organisms.

  E. Biomass and Ecological Pyramids

    1. Dry organic matter produced by living things is termed biomass. Little biomass, usually 5-20%, actually passes from one trophic level to the next. As a result, graphic representations of ecosystems show pyramids of biomass. energy, or numbers. These pyramids have important ecological and human implications.

    2. Eating lower in the food chain (grains instead of meat) have advantages for food production since 80-90% of the available energy is lost at each trophic level.

  F. Nutrient Cycles - Nutrients move through the biosphere in biogeochemical cycles. Each cycle involves an environmental and an organismic phase.

    1. The Carbon Cycle - Photosynthesis and cellular respiration (an oxidative process) cycle carbon and oxygen in the biosphere. Humans affect the carbon cycle by deforestation (reducing photosynthesis) and burning (oxidizing) fossil fuels.

    2. The Nitrogen Cycle - Atmospheric nitrogen is made available to plants by a variety of organisms capable of nitrogen fixation. Modern farming practices are interfering with the natural cycling of nitrogen.

3. The Phosphorus Cycle - Phosphorus-rich rocks slowly release phosphates to soils and water. Artificial fertilizers replace phosphates lost to runoff; this practice often causes pollution problems in nearby aquatic systems.

Suggestions for Presenting the Chapter

- An examination of local endangered species and their habitat needs may illustrate the impact of humans on natural systems. Focus can be placed on the abiotic and biotic factors necessary for the survival of a particular endangered species.
- A field trip to explore local aquatic habitats (marshes, lakes, streams, etc.) will aid in understanding of aquatic life zones. An exercise focusing on the difference in habitat types and the adaptations necessary for life in these habitats is useful in illustration of several concepts. Food chains/webs, trophic levels and the concept of niche can all be explored during this activity.
- Instructors might supplement their lectures with videos focusing on biomes and aquatic habitats. Students can also be assigned video viewing outside of class time in conjunction with assigned readings or worksheets.
- Students can be asked to identify threatened ecosystems. Many fine articles are found in common environmental periodicals (Audubon, National Wildlife, etc.).
- Recommended website:

The Virtual Library of Ecology and Biodiversity: http://conbio.rice.edu/vl/
Ecology WWW Page: http://pbil.univ-lyon1.fr/Ecology/Ecology-WWW.html
The World's Biomes: http://www.ucmp.berkeley.edu/glossary/gloss5/biome/
Major Biomes of the World:
http://www.runet.edu/~swoodwar/CLASSES/GEOG235/biomes/main.html

Ecological Society of America/Educational Resources: http://esa.sdsc.edu/ed_resources.htm

True/False Questions

1. ____ Ecology is the study of organisms and their relationship to their environment.
2. ____ The word "ecology" is synonymous with the word "environment".
3. ____ The biosphere consists of distinct regions called biomes and aquatic life zones.
4. ____ The hydrosphere contain the gases found in the air surrounding the earth.
5. ____ The lithosphere is comprised of all the material contained in the earth.
6. ____ Plants use sunlight, air and soil nutrients in the process of photosynthesis to produce new molecules and tissues.
7. ____ The biosphere extends from the bottom of the ocean to the tops of mountains.
8. ____ The biosphere is an open system.
9. ____ A biome is a terrestrial region with a distinct assemblage of plants and animals.
10. ____ The taiga contains permafrost and no trees.
11. ____ The taiga is also called the boreal forest.
12. ____ Cacti, mesquite trees, lizards and rattlesnakes are common members of the desert biome.

13.___ The grassland biome contains deep rooted grasses than can withstand drought and frequent grazing.

14.___ Grazer food chains begin with herbivores being eaten by carnivores.

15.___ Decomposer food chains start with dead material.

16.___ Phytoplankton are small, microscopic photosynthesizers that are found primarily in the atmosphere.

17.___ Biomass is the dry weight of living material in an ecosystem.

18.___ Herbivores are part of the first trophic level in the grazer food chain.

19.___ The carbon cycle is being flooded with excess carbon dioxide as a result of the combustion of fossil fuels and deforestation.

20.___ Nitrogen fixation occurs when carbon dioxide is taken out the air by plants and bacteria.

Fill-in-the-Blank Questions

1. Nitrogen, like _____ is a plant nutrient.

2. The roots of _____ plants contain root nodules that are involved in nitrogen fixation.

3. In the past 100 years, global atmospheric carbon dioxide levels have increased _____%.

4. Today, __ billion tons of carbon dioxide are added to the atmosphere each year.

5. The term _____ refers to all ions and molecules used by living organisms.

6. Producers are on the _____ trophic level.

7. _____ is the dry weight of living material in a ecosystem.

8. A _____ is a series of organisms, each feeding on the organism preceding it.

9. Grazer food chains begin with plants and _____.

10. _____ food chains begin with dead material.

11. _____ are organisms such as algae and plants that absorb sunlight and use its energy to make biomass.

12. _____ are organisms that feed on animal waste or the remains of plants and animals.

13. Animals that feed on herbivores and other animals are called _____.

14. The place that an organism normally lives is called its _____.

15. The oceans can be divided into distinct zones called _____ life zones.

16. The range of conditions to which an organism is adapted is called its range of _____.

17. East of the Mississippi River the _____ deciduous forest is the primary biome.

18. The biosphere is a _____ system.

19. The largest biological system is the _____.

20. A _____ is a terrestrial region characterized by a distinct assemblage of plants and animals.

Fill-in Key:

1. phosphorus
2. leguminous
3. 100
4. 7
5. nutrients
6. first
7. biomass
8. food chain
9. algae
10. decomposer
11. producers
12. detritivores
13. carnivores
14. habitat
15. aquatic
16. tolerance
17. temperate
18. closed
19. biosphere
20. biome

Multiple Choice Questions

1. The science that studies the relationship between organisms and the environment is:
   a. geology
   b. biology
   * c. ecology
   d. physiology
   e. anthropology

2. The biosphere is a _____system consisting of air, water and soil that are recycled over and over.
   a. open
   * b. closed
   c. static
   d. nonrenewable
   e. economic

3. The largest biological system is the:
   a. atmosphere
   b. lithosphere
   c. ionosphere
   * d. biosphere
   e. stratosphere

4. The process where plants use sunlight to produce carbohydrates from carbon dioxide is called:
   * a. photosynthesis
   b. respiration
   c. nitrogen fixation
   d. ATP production
   e. biological oxidation

5. Because the earth's ecological systems are closed, all materials necessary for life must be:
   a. synthetic
   b. organic
 * c. recycled
   d. decomposed
   e. fermented

6. Terrestrial areas on earth containing specific chemical and physical conditions allowing survival of a specific assemblage of organisms is called a:
   a. ecosystem
   b. biosphere   d. niche
 * c. biome   e. cline

7. The aquatic equivalent of a biome is :
   a. rhizome
 * b. life zone
   c. ocean
   d. zone of tolerance
   e. niche

8. The range of conditions to which an organism is adapted is the:
   a. niche
   b. biome
 * c. range of tolerance
   d. zone of intolerance
   e. limiting factor

9. The area where organisms cannot survive exposure to environmental conditions is the:
   a. range of tolerance
 * b. life zone b. zone of intolerance
   c. zone of physiological stress
   d. zootic climax
   e. taiga

10. The primary competitive advantage our species has over every other lies in our use of:
   a. biological control agents
   b. organic farming
   c. sustainable energy sources
 * d. technology
   e. efficient energy transformation

11. The northern coniferous forest is also called the:
   a. taiga
 * b. tundra
   c. barren ground
   d. temperate forest
   e. savanna

12. This biome contains grasses, mosses, lichens, wolves, musk oxen and arctic fox:
   a. chaparral
   b. taiga
 * c. tundra
   d. boreal forest
   e. deciduous forest

13. In freshwater lakes and rivers, is a limiting factor to primary productivity.
   a. carbon dioxide
 * b. phosphate
   c. carbonate
   d. oxygen
   e. water

14. Organisms of the same species within a biome or aquatic life zone are called a:
   a. community
 * b. population
   c. niche
   d. food web
   e. biome

15. It is estimated that 40-100 species become extinct every day largely because of:
   a. aquatic pollution
   b. destruction of tundra
   c. desertification
 * d. tropical deforestation
   e. ozone depletion

16. A complex network of feeding interactions in a biological community is a:
   a. food chain
 * b. food web
   c. niche
   d. trophic level
   e. food pyramid

17. This group of organisms use sunlight to synthesize organic molecules and thereby provide nourishment for all other organisms?
   a. consumers
   b. decomposers   d. herbivores
   c. detritivores   *e. producers

18. This group of organisms feeds on the plants produced on earth:
   a. predators
* b. herbivores
   c. autotrophs
   d. producers
   e. carnivores

19. A series of organisms, each one feeding on the preceding organism forms a:
   a. niche
   b. habitat
   c. biome
* d. food chain
   e. nutrient cycle

20. These food chains begin with dead material:
   a. grazer food chains
   b. predator food chains
   c. herbivore food chains
* d. decomposer food chains
   e. aquatic grazer

21. Coyotes, wolves and mountain lions are meat eaters and would be called:
   a. herbivores
   b. producers
   c. autotrophs
* d. carnivores
   e. omnivores

22. Excessive levels of ultra violet radiation reaching earth are associated with increased risk of cancer in humans. UV radiation is also an environmental concern since it can kill _____ and cause collapse of food chains.
   a. rodents
   b. frogs
* c. phytoplankton
   d. predators
   e. carnivores

23. The position of an organism in a food chain is called a:
   a. niche
   b. food web
* c. trophic level
   d. ecotone
   e. optimum range

24. The nonliving components of an ecosystem are called:
   a. life zones
   b. biotic
* c. abiotic
   d. taiga
   e. decomposition

25. Two species cannot occupy the same niche for long periods. This principle is called:
   a. range of tolerance
   b. limiting factor
   c. zone of intolerance
   d. life zone
* e. competitive exclusion

26. Populations of elk and mule deer are greatly influenced by the availability of good winter range habitat. Winter range would be considered to be a _____ to these populations.
   a. niche
 * b. limiting factor
   c. trophic level
   d. ecotone
   e. biome

27. Plants belong to the:
* a. first trophic level
   b. second trophic level
   c. third trophic level
   d. fourth trophic level
   e. fifth trophic level

28. Herbivores belong to the:
   a. first trophic level
* b. second trophic level
   c. third trophic level
   d. fourth trophic level
   e. fifth trophic level

29. The dry weight of living material in the ecosystem is called:
    a. detritus
 * b. biomass
    c. biota
    d. biosphere
    e. biome

30. An organism's relationships to the and biotic components of the environment is called the:
    a. habitat
 * b. niche
    c. ecotone
    d. biome
    e. ecosystem

31. The permanent loss of a species from the biosphere is called:
    a. relocation
    b. migration
    c. emigration
    d. adaptation
 * e. extinction

32. The dodo and the passenger pigeon were exterminated by:
 * a. overhunting
    b. poisoning
    c. habitat destruction
    d. disease
    e. climate change

33. Rampant deforestation and use of fossil fuels have overloaded the carbon cycle with which of the following substances?
    a. nitrate
    b. phosphate
    c. water
 * d. carbon dioxide
    e. nitrogen

34. In the past 100 years the global atmospheric carbon dioxide levels have increased:
    a. 10%
    b. 15%
 * c. 25%
    d. 50%
    e. 75%

35. The conversion of nitrogen to ammonia is called:
    a. eutrophication
    b. decomposition
 * c. nitrogen fixation
    d. denitrification
    e. acidification

36. Nitrogen oxides released by power plants and automobiles may be converted to _____ which alters the pH of the environment.
    a. sulfuric acid
 * b. nitric acid
    c. amino acids
    d. nitrogen gas
    e. DNA

37. Which of the following organisms would be tertiary consumers?
    a. grasses
    b. trees
    c. mushrooms
    d. insects
 * e. hawks

38. In the carbon cycle, animal respiration returns _____ to the atmosphere.
    a. nitrogen
    b. oxygen
 * c. carbon dioxide
    d. carbonic acid
    e. sulfuric acid

39. Human application of anhydrous ammonia to the soil as a fertilizer would be expected to increase the levels of soil _____ which can be used by plants as a nutrient.
    a. bacteria
    b. nitrous oxide
 * c. nitrates
    d. free nitrogen gas
    e. phosphates

40. This substance can be converted by cyanobacteria in the soil to ammonia?
    a. carbon dioxide
    b. oxygen
 * c. nitrogen
    d. water
    e. methane

43

6
Principles of Ecology: Biomes and Aquatic Life Zones

Chapter Outline

Weather and Climate: An Introduction
    What factors influence weather and climate?
    The Coriolis Effect
    Ocean Currents

The Biomes
    Tundra
    Taiga
    Temperate Deciduous Forest
    Grassland
    Desert
    Tropical Rain Forest
    Altitudinal Biomes

Aquatic Life Zones
    Freshwater Lakes
    Rivers and Streams
    Protecting Freshwater Ecosystems
    Saltwater Life Zones

Key Terms

weather	polar easterlies	grassland
climate	Gulf Stream	short-grass prairie
trade winds	biomes	tall grass prairie
tropics	tundra	dust bowl
temperate zone	taiga	lateritic soils
polar regions	temperate deciduous forest	altitudinal biomes
Coriolis Effect	desert	alpine tundra
westerlies	tropical rain forest	aquatic life zones

phytoplankton	zooplankton	freshwater lakes
littoral zone	limnetic zone	profundal zone
benthic zone	epilimnion	thermocline
hypolimnion	fall overturn	spring overturn
watershed	estuary	coastal wetlands
coral reef	continental shelf	continental slope
abyssal plain	neritic zone	euphotic zone
abyssal zone	bathyal zone	

Objectives

1. Discuss the major factors that determine weather and climate.
2. Explain the Coriolis Effect and the resulting global air circulation and wind patterns.
3. List the earth's major biomes and their characteristics.
4. Discuss the major freshwater aquatic life zones.
5. Describe the major saltwater life zones.
6. Discuss the four life zones of the marine ecosystem.
7. Explain why the world's richest biome grows on some of the world's poorest soils.
8. Explain why there is a yearly cycle of "turnover" in many freshwater lakes.
9. Summarize some adaptations of organisms to different biomes/aquatic life zones.
10. Discuss the phenomenon called "El Niño".

Lecture Outline

I. Weather and Climate: An Introduction
   A. *Weather* refers to the daily conditions in our surroundings, including temperature and rainfall. Climate is the average weather over a along period, approximately 30 years. The *climate* of a region determines what plants can live there.
   B. Major Factors That Determine Weather and Climate
      1. The earth is unequally heated creating three major climatic zones:
         a. Tropical - between 30° north and 30° south latitudes.
         b. Temperate - between 30° and 60° of both hemispheres.
         c. Polar - above 60° in both hemispheres.
      2. The temperature and rainfall patterns in the above zones are also influenced by the flow of moist air and heat from the equator.
      3. When air is heated it expands and becomes less dense. It rises and moves toward the poles. Cold air from the poles moves toward the temperate zones in each hemisphere.
   C. The Coriolis Effect - The deflection of wind currents by the spin of the earth on its axis.
      1. The following wind patterns are formed in the major climatic zones listed above:
         a. Trade winds - These winds are found in the tropics and are easterlies.
         b. Westerlies - These winds are found in temperate regions.
         c. Polar Easterlies - These winds are found in polar regions.
   D. Ocean Currents - Warm water from the equator flows toward the poles, warming land masses near which it passes.

1. Gulf Stream - a current that flows from the equator northward carrying warmer water.
2. Humboldt current - a current that flows from the southern polar regions toward the equator along the western side of South America. This current is a deep current emerges at the surface in regions called *upwellings*. Every 10 years or so, the Humboldt current turns warm causing the phenomenon called *El Niño*. This weather pattern creates heavy rain and rough seas and can have a devastating effect on coastal communities.

II. The Biomes - Biomes are regions with a distinct climate and a unique assemblage of plants and animals.
   A. The Tundra - Tundra is the biome characterized by the harshest climate.
      1. Found in the northernmost portions of North America, Europe and Asia.
      2. Tundra covers about 10% of the earth.
      3. Tundra is a treeless area characterized by grasses, shrubs, and mat-like vegetation (mosses and lichens).
      4. Tundra receives less that 25 centimeters or 10 inches of precipitation per year.
      5. The deeper layers of the soil remain frozen throughout the year and are called the *permafrost*.
      6. Summer on the tundra is a time of great activity. Many birds nest in the tundra and take advantage of the swarms of insects for food to rear their young. Ptarmigan, musk oxen, arctic hares, arctic foxes, wolves and polar bears are all adapted to survive on the tundra.
   B. The Taiga - The taiga is a band of coniferous forest spreading across the northern continents just below the tundra.
      1. Sometimes called the *northern coniferous biome*, this land of pine, fir and spruce trees extends across Canada, part of Europe, and Asia.
      2. The taiga receives more precipitation than the Tundra and is able to support deep-rooted plants such as trees (38-100cm/15-40in)
      3. The forests and other habitats of the taiga support a variety of species such as bear, moose, deer, and many smaller mammals.
      4. The forests of the taiga are being harvested for their timber at a high rate. There is much concern for the old-growth forests of the taiga since few remain undisturbed.
   C. The Temperate Deciduous Forest Biome - This biome occurs in regions with abundant rainfall and long growing seasons (75-150cm/30-60in).
      1. The temperate forest biome is located in the eastern United States, Europe, and northeast China. In the U.S. this biome is home to about half of the human population.
      2. The dominant plants of the temperate deciduous forest are deciduous trees, the broad-leaved trees that shed their leaves each fall. Maple, oak, black cherry, and beech tress are examples of some deciduous trees.
      3. The temperate forest has a deep, rich soil that supports numerous crops.
      4. The fertile soil and abundant plant life of the forest support a rich and varied population of insects, microorganisms, birds, reptiles, amphibians and mammals.
      5. This biome has been greatly altered by human activities and much of the original forest area is no longer remaining. East of the Mississippi river, only 10% of the land is still forested and only 0.1% of the original forest remains. Most of the forests are second, third, or fourth growth stands.

D. The Grassland Biome - The grassland biome occurs in regions of intermediate precipitation, enough to support grasses but not enough to support trees (25-75cm/10-30in).
  1. Grasslands exist in temperate and tropical regions. Grasslands are found in North America, South America, Africa, Europe, Asia, and Australia.
  2. All grasslands bear a similar topography; most are on flat or slightly rolling terrain.
  3. The soils found in grasslands are among the richest in the world. As a result, grasslands have been exploited for growing crops.
  4. Grasslands in North America once supported an enormous population of bison, elk, and even grizzly bears.
  5. Poor agricultural practices on farms and ranches throughout the world have resulted in widespread soil erosion and desertification in the grassland biome. The most dramatic and well know example of abuse came in the 1930s on the great plains. Millions of tons of topsoil were lost from farms in the U.S. during the *dust bowl*.

E. The Desert Biome - The desert biome is characterized by dry, hot conditions but often abounds with plants and animals adapted to the heat and lack of rainfall.
  1. Deserts exist throughout the world. The Sahara stretches across northern Africa and is about the size of the United States. Deserts in North America exist primarily on the downwind side of the mountain ranges.
  2. Precipitation in deserts is low (under 25cm/10in).
  3. Plants living in desert conditions are adapted to low soil moisture and can survive extreme temperature changes. Water is stored by many desert plants in succulent, water-retaining tissues, giving the plant an ample supply during the rainless months.
  4. Many insects and other animals make the desert their home. The thick scales of snakes and lizards is an adaptation to reduce water loss.
  5. Large cities have sprung up in many of the world's deserts. This can create serious water problems as a result of depletion of ground and surface waters.
  6. Each year, millions of acres of new desert form on semiarid grasslands. Deserts are expanding primarily because of human actions. Livestock overgrazing destroys grasses and can remove enough vegetation to reduce rainfall. This process of expanding deserts is called *desertification*.

F. The Tropical Rain Forest Biome - This biome is the richest and most diverse biologically.
  1. Tropical rain forest receives abundant rainfall (150-400cm/60-160in) and has a warm climate
  2. Tropical rain forests exist near the equator in South and Central America, Africa, and Asia.
  3. This is the most diverse of all biomes and supports an incredible variety of life.
  4. The tropics actually contain a variety of different ecosystems of which the rain forest is the largest and best known.
  5. Tropical forest soils are poor in nutrients. The biomass that reaches the ground is quickly used by forest organisms and is not accumulating as it does in the temperate deciduous forest. Nutrients released into the soil by decomposers is quickly absorbed by the roots of trees. The nutrients are very quickly cycled back into the plants and animals of the forest.
  6. Many rainforest soils are useless for long-term farming and ranching since they are high in iron content (laterite or lateritic soils; meaning "brick"). When

cleared and exposed to sunlight, lateritic soils bake as hard as bricks making it difficult to cultivate.

7. Tropical trees are the lungs of the planet. They absorb carbon dioxide and release oxygen during daytime hours. The absorption of carbon dioxide reduces global warming.

8. Rainforests may be the most threatened biome on earth. A recent United Nations study suggested that about 17 million hectares (42 million acres) were being cleared each year. This is an area about the size of the state of Washington. Widespread destruction of tropical forests could upset global rainfall patterns and could affect global climate by reducing the amount of carbon dioxide removed from the atmosphere each year.

G. Altitudinal Biomes - Because climate varies with altitude, the distribution and abundance of life also changes.

1. Altitudinal biomes mirror the latitudinal biomes discussed earlier. These result from differences in precipitation and temperature associated with altitude.

III. Aquatic Life Zones - The aquatic equivalent of biomes on land; these regions of life are influenced by available energy and nutrients.

A. Freshwater Lakes -

1. Lakes are divided into four habitats:
   a. Littoral zone - The littoral zone consists of shallow waters at the margins of a lake. Aquatic vegetation is often growing in this area.
   b. Limnetic zone - This area extends from the edge of the littoral zone to the point of depth were light no longer penetrates. This is the main photosynthetic zone of the lake and supports abundant phytoplankton, the primary producers of the lake ecosystem.
   c. Produndal zone - This region lies beneath the limnetic zone in deeper lakes and extends to the bottom. Little or no light enters this area and conditions are not favorable for plant and algal growth. Fishes can survive in this region but rely on food produced in the limnetic an littoral zones.
   d. Benthic zone - The bottom of the lake is the benthic zone. Organisms such as snails, clams, crayfish, aquatic worms, and insect larvae live in bottcm sediments.

2. Temperature layers - Deeper lakes often contain three distinct temperature layers during the summer months. Twice a year the lake waters will turnover mixing the surface and deep water (spring and fall overturn). The summer layers of the lake are listed below:
   a. Epilimnion - This is the uppermost layer of the lake containing warm water.
   b. Thermocline - This is a layer of water where the temperature gradient is greater than in the epilimnion or hypolimnion.
   c. Hypolimnion - This is the layer of water below the thermocline and is the coldest layer.

B. Rivers and Streams - Rivers and streams are complex ecosystems that rely more on agitation for oxygenation of their water than lakes do.

1. Watershed - A watershed is the region drained by a stream.

2. Streams may receive much of their nutrients from the surrounding terrestrial ecosystems. Leaves, animal feces, seeds, stems and other biomass may wash or fall into streams.

3. Many primary consumers in streams are detritivores.

4. Primary productivity varies in streams and depends on the amount of light reaching the water. Producers like algae and mosses and rooted vegetation are found in stream ecosystems.

C. Protecting Freshwater Ecosystems - Lake are repositories for pollutants and are highly vulnerable to them. Streams generally fare better than lakes because of their flow, which tends to whisk pollutants away.

1. Protecting lakes, rivers and streams requires measures to control or eliminate pollution.

2. Waste reduction is a more sustainable strategy. Companies are learning that if they can eliminate the use of toxic substances this is often cheaper than end-of-pipe controls.

D. Saltwater Life Zones - The oceans of the world consist of distinct life zones differing in biotic and abiotic conditions.

1. Coastal life zones - Coastlines are highly productive waters characterized by abundant sunlight and a rich supply of nutrients, which contribute to an abundance of life-forms.

   a. Estuaries and Coastal Wetlands - Estuaries are areas where freshwater mixes with salt water. Estuaries are rich in life because streams and rivers transport nutrients from the land and incoming tides carry nutrients into estuaries from the sea. Coastal wetlands (salt marshes, mangrove swamps, and mud flats) are located near estuaries. Human activities that disrupt rivers like dams can have a devastating effect on these ecosystems.

   b. The Shoreline - Shorelines are rocky or sandy regions that are home to a surprising variety of organisms adapted to the tide and to the turbulence created by wave action

   c. Coral Reefs - Coral reefs are wonderful areas of high diversity, the equivalent of the tropical rainforest in water! Coral reefs are found in relatively warm and shallow waters in the tropics or subtropics.

2. The Marine Ecosystem - The marine ecosystem consists of four ecologically distinct life zones:

   a. Neritic zone - The neritic zone is similar to the littoral zone of lakes. This zone lies above the continental shelf and contains relatively shallow water and receives abundant light. Most commercial fishing is done in the neritic zone.

   b. Euphotic zone - This zone is similar to the limnetic zone of lakes. It is the open-water region that extends to the lower limits of sunlight (about 200 meters; 650 feet). This region supports numerous species of phytoplankton.

   c. Bathyal zone - This zone is a region of semidarkness and is too dark to support photosynthesis. The bathyal zone is home to a variety of animals which feed on material sinking in the water column

   d. Abyssal zone - This zone is a region of complete darkness. It has no photosynthetic organisms and is low in oxygen. Animals that live in this zone are adapted to cold water, high pressure, low oxygen and complete darkness. Sediment in the abyssal zone is often rich in nutrients.

Suggestions for Presenting the Chapter

- Nothing is better for students than seeing biomes in person. Fieldtrips to local biomes is an excellent way to introduce the concepts in this chapter.
- If fieldtrips are not practical in your setting there are many excellent environmental videos that examine biomes and ecosystems.
- Classes can be involved in monitoring local weather conditions, perhaps, in conjunction with local TV meteorologists or meteorology classes on campus.
- Worksheets can be prepared to examine local climatic conditions such as mean annual temperature, monthly, seasonal and yearly precipitation, and even the effects of El Niño.
- Recommended websites:

  National Oceanographic and Atmospheric Administration: http://www.lib.noaa.gov/
  Marine Biology Web:http://life.bio.sunysb.edu/marinebio/mbweb.html
  Environmental Journals on the Internet: http://www.cnie.org/Journals.htm
  Oceans and Coastal Resources: A Briefing Book/NCSE: http://www.cnie.org/nle/mar-20/
  El Niño and Southern Oscillation (ENSO) Homepage: http://www.ogp.noaa.gov/enso/

True/False Questions

1. ___ Weather refers to the daily conditions in our surroundings including temperature and rainfall.
2. ___ Climate is the average weather over a long period, approximately 30 years.
3. ___ The earth is equally heated by the sun in all major climatic regions.
4. ___ The tropics lie on either side of the equator between 30 degrees north and 30 degrees south latitude.
5. ___ The winds that blow toward the equator are called the westerlies.
6. ___ The Humboldt current is found in the Atlantic ocean off the east coast of the United States.
7. ___ Every 10 years the Humboldt current turns warm, a phenomenon that is called El Niño.
8. ___ The northern most biome is treeless and has permafrost and is called arctic tundra.
9. ___ The taiga is also called the northern coniferous forest.
10. ___ The dominant trees in the deciduous forest are conifers.
11. ___ The grassland biome receives 10-30 inches of precipitation.
12. ___ Poor agricultural practices in the 1930s produced the "dust bowl" in midwestern and western states.
13. ___ Lateritic soils are found in temperate grasslands and forests.
14. ___ The tropical rain forest is the richest and most diverse biome on earth.
15. ___ The warm surface water of a lake is called the thermocline.
16. ___ The bottom of the lake is called the limnetic zone.
17. ___ The littoral zone consists of the shallow waters at the margin of the lake.
18. ___ The region drained by a stream is called a watershed.
19. ___ The gradually sloping region of the ocean floor is called the continental slope.
20. ___ The neritic zone of the ocean is equivalent to the littoral zone of lakes.

Fill-in-the-Blank Questions:

1. The _____ shelf is the gradually sloping portion of the sea floor.
2. The bottom of the deep ocean is called the _____ plain.
3. _____ reefs are the aquatic equivalent of the tropical rain forest.
4. The _____ zone is the oceanic equivalent of the limnetic zone of lakes.
5. _____ are places where freshwater mixes with salt water.
6. The warm surface water of a lake is the _____ .
7. The region drained by a stream is called a _____ .
8. The mixing of surface and bottom waters of lakes is know as the fall _____ .
9. The producers in freshwater and saltwater life zones are the _____ .
10. The treeless biome found on the top of mountains that is similar to arctic tundra is _____ tundra.
11. The dominant plants of the temperate deciduous forest biome are _____ trees.
12. A wide band of coniferous trees found in Canada, Europe and Asia is called the

    _____ .
13. _____ have narrow, pointed leaves called needles.
14. In the arctic, deeper layers of soil remain frozen throughout the year and are called

    _____ .
15. The deflection of wind currents by the spin of the Earth is called the _____ effect.
16. The cold Arctic air that blows from the northeast to southwest is called the _____ easterlies.
17. _____ is the average weather recorded over a period of 30 years.
18. _____ is daily conditions such as rainfall and temperature.
19. The Earth is unequally heated which creates three major _____ zones.
20. The earth can be divided into _____ climate zones.

13. conifers
14. permafrost
15. coriolis
16. polar
17. climate
18. Weather
19. Climatic
20. Three

Multiple Choice Questions

1. The daily conditions in our surroundings, including temperature and rainfall is called:
   a. climate
   * b. weather
   c. biome
   d. climate zone
   e. zonation

2. The average weather over a long period is called:
   * a. climate
   b. weather
   c. biome
   d. climate zone
   e. zonation

3. The three major climatic zones are:
   a. tropical, subtropical and arctic
   b. tropical, subarctic, and arctic
   * c. tropical, temperate, and polar
   d. subtropical, tropical, and temperate
   e. subtropical, temperate and arctic

4. The deflection of wind currents by the rotation of the earth results from the:
   a. topography
   b. El Niño
   c. upwellings
   * d. Coriolis Effect
   e. Rain Shadow Effect

5. The cold Humboldt current flows toward the equator and emerges at the surface of the ocean in regions called:
   a. life zones
   b. sinks
   c. tidal pools
   d. breakers
   * e. upwellings

6. Terrestrial areas with a distinct climate and assemblage of plants and animals is called a:
   * a. biome
   b. ecosystem
   c. biosphere
   d. ecotone
   e. life zone

7. The northernmost biome is characterized by permafrost and is treeless. It is called the:
   a. temperate deciduous forest
   b. desert
   c. taiga
   * d. tundra
   e. grassland

8. The taiga is another name for the _____ biome.
   a. tundra
   * b. northern coniferous forest
   c. desert
   d. tropical rain forest
   e. grassland

9. The dominant plant type of the taiga are:
    a. grasses
    b. deciduous trees
*   c. conifers
    d. scrubs
    e. epiphytes

10. Deciduous trees lose their leaves in the fall
    which is an adaptation to:
    a. hot weather
    b. cold weather
*   c. prevent water loss
    d. reduce energy expenditure
    e. increase internal metabolism

11. The tall-grass and short-grass prairie habitats
    are a good example of the _____ biome.
    a. temperate deciduous forest
    b. desert
*   c. grassland
    d. tropical rain forest
    e. tundra

12. This biome has under 25 centimeters
    (10inches) of precipitation each year and
    abounds with plants and animals adapted to
    heat.
*   a. desert
    b. temperate deciduous forest
    c. grassland
    d. tropical rainforest
    e. taiga

13. Water produced by cellular energy
    production in animals is called _____
    water.
    a. internal
    b. cellular
    c. conserved
    d. free
*   e. metabolic

14. This biome receives 200-400 centimeters(60-
    160 inches) of precipitation each year and
    has the highest species diversity of any
    biome.
    a. desert
    b. deciduous forest
    c. taiga
*   d. tropical rain forest
    e. tundra

15. The aquatic equivalent of the terrestrial biome
    is the:
    a. ecotone
    b. estuary
    c. tidal zone
*   d. aquatic life zone
    e. altitudinal biome

16. Microscopic producers in aquatic ecosystems
    are called:
    a. zooplankton
*   b. phytoplankton
    c. krill
    d. kelp
    e. lichens

17. Which of the following is one of the regions
    of a freshwater lake?
*   a. littoral zone
    b. estuarine zone
    c. euphotic zone
    d. abyssal plain
    e. continental shelf

18. The three temperature layers in a freshwater
    lake during the summer months are the:
    a. epilimnion, euphotic zone, & hypolimnion
    b. profundal zone, benthic zone, and shore
*   c. epilimnion, thermocline, and hypolimnion
    d. hypolimnion, thermocline, and abyss
    e. photic zone, euphotic zone and neritic
       zone

19. The oceans cover ____ percent of the earth's
    surface.
    a. 15
    b. 35
    c. 50
*   d. 70
    e. 85

20. These saltwater habitats are located at the
    mouths of rivers where fresh and saltwater
    mix and are often associated with coastal
    wetlands:
    a. coral reefs
*   b. estuaries
    c. tidal pools
    d. continental shelf
    e. abyssal plain

21. Salt marshes, mangrove swamps and mud flats are all:
   a. coral reefs
   b. tidal pools
   c. estuaries
   * d. coastal wetlands
   e. shorelines

22. Studies show that approximately two-thirds of the world's commercially valuable fishes and molluscs depend on this aquatic life zone at some point in their life cycle:
   a. tidal pools
   b. rivers
   * c. estuaries
   d. coastal wetlands
   e. shorelines

23. This aquatic life zone is found in saltwater and consists of calcium carbonate or limestone deposits. This life zone is the aquatic equivalent of the tropical rain forest:
   a. shorelines
   b. estuaries
   * c. coral reefs
   d. salt marshes
   e. coastal wetlands

24. The region of the sea floor that extends from the land masses and slopes gradually is called the:
   a. abyssal plain
   b. shoreline
   * c. continental shelf
   d. continental slope
   e. euphotic zone

25. The bottom of the deep ocean is called the:
   * a. abyssal plain
   b. shoreline
   c. continental shelf
   d. continental slope
   e. euphotic zone

26. This zone of the marine ecosystem contains relatively shallow water, is warm and well oxygenated. It is the equivalent to the littoral zone of lakes:
   * a. neritic zone
   b. euphotic zone
   c. abyssal zone
   d. bathyal zone
   e. photic zone

27. This zone of the marine ecosystem is the oceanic equivalent of the limnetic zone of lakes:
   a. neritic zone
   * b. euphotic zone
   c. abyssal zone
   d. bathyal zone
   e. littoral zone

28. This zone of the marine ecosystem is too dark to support photosynthesis and has low oxygen levels. This region of semidarkness is know as the:
   a. neritic zone
   b. euphotic zone
   c. abyssal zone
   * d. bathyal zone
   e. littoral zone

29. This zone of the marine ecosystem is a region of complete darkness along the deep ocean floor and contains no photosynthetic organisms:
   * a. abyssal zone
   b. neritic zone
   c. euphotic zone
   d. bathyal zone
   e. littoral zone

30. The climate zone on either side of the equator between 30° north and 30° south latitude is the:
   * a. tropics
   b. temperate zone
   c. polar zone
   d. arctic zone
   e. taiga

31. These winds blow from northeast to southwest in the tropics, they are the:
    a. westerlies
    b. polar easterlies
*   c. tradewinds
    d. jet stream
    e. chinooks

32. About every 10 years the Humboldt current turns warm and creates heavy rain, rough seas and climatic changes. This weather pattern is called:
    a. jet stream
    b. global warming
    c. rain shadow effect
*   d. El Niño
    e. Coriolis Effect

33. Ptarmigan, musk oxen, arctic hares and arctic fox would be animals found in which of the following biomes?
*   a. arctic tundra
    b. temperate deciduous forest
    c. grasslands
    d. desert
    e. tropical rainforest

34. In the United States only about ___ % of the land east of the Mississippi River is still forested.
*   a. 10
    b. 20
    c. 30
    d. 40
    e. 50

35. In grassland biomes, which of the following human activities is responsible for the greatest environmental impact?
    a. oil production
    b. coal mining
    c. sport hunting and recreation
*   d. farming
    e. mineral extraction

36. Sagebrush becomes a dominant member of the community as a result of overgrazing which of the following habitats?
    a. tall grass prairie
*   b. short-grass prairie
    c. sand prairie
    d. hill prairie
    e. temperate deciduous forest

37. Which of the following are adaptations found in desert plants?
    a. deep taproots
    b. thick, waxy coverings
    c. thorns
    d. water-retaining tissues
*   e. all of the above

38. Lateritic soils would be would in which of the following biomes?
    a. arctic tundra
    b. taiga
    c. temperate deciduous forest
    d. grasslands
*   e. tropical rain forest

39. Alpine tundra would be a good example of a/an:
    a. grassland
    b. climax ecosystem
*   c. altitudinal biome
    d. aquatic life zone
    e. microhabitat

40. This zone in freshwater lakes lies beneath the limnetic zone and extends to the bottom?
*   a. profundal zone
    b. benthic zone
    c. photic zone
    d. abyssal zone
    e. neritic zone

7

Principles of Ecology: Self-Sustaining Mechanisms in Ecosystems

Chapter Outline

Homeostasis: Maintaining the Balance
Homeostasis in Natural Systems
What factors contribute to ecosystem homeostasis?
The Resilience of Ecosystems
Resisting Changes From Human Activities
Population Control and Sustainability
Species Diversity and Stability

Natural Succession: Establishing Life on Lifeless Ground
Primary Succession
Secondary Succession
Changes During Succession: An Overview

Evolution: Responding to Change
Evolution by Natural Selection
Genetic Variation
Natural Selection
Speciation
Coevolution and Ecosystems Balance

Human Impact on Ecosystems
Tampering with Abiotic Factors
Tampering with Biotic Factors
Simplifying Ecosystems
Why Study Impacts?

Key Terms

Homeostasis
environmental resistance
primary succession
climax communities
natural selection
crossing over
speciation
sympatric speciation
pathogens
mitigate

growth factors
species diversity
pioneer communities
secondary succession
adaptations
selective advantage
reproductive isolation
coevolution
monoculture
Environmental Impact Statement

reduction factors
natural succession
intermediate communities
evolution
gametes
fitness
allopatric speciation
restoration ecology
ecosystem simplification

Objectives

1. Define "homeostasis" and discuss how this term relates to natural systems.
2. Discuss the factors that contribute to ecosystem homeostasis.
3. List the mechanisms of population regulation seen in natural systems.
4. Discuss how species diversity can affect ecosystem stability.
5. Define "natural succession", "primary succession" and "secondary succession".
6. Discuss the changes that occur during the process during natural succession.
7. Define "evolution", "natural selection" and "adaptations".
8. List the sources of genetic variation that can be acted on by natural selection.
9. Define "speciation", "allopatric speciation", and "sympatric speciation".
10. Define "coevolution" and give an example of how it works.
11. Define "restoration ecology" and "conservation biology".
12. Discuss the potential impact caused by simplifying ecosystems.
13. Discuss the significance of the "Environmental Impact Statement" for predicting the affect of human activities.

Lecture Outline

I. Homeostasis: Maintaining the Balance - Homeostasis in an ecosystem is a steady state or dynamic equilibrium, where internal conditions are held more or less constant despite changes in external conditions.
    A. Homeostasis in Natural Systems - Ecosystems have many mechanisms that resist change or help them recover from change which helps to keep natural systems in balance.
        1. Predators - Predators reduce the population size of the prey they feed upon.
        2. Diseases - Disease is also a mortality factor that can stabilize population size.
    B. What Factors Contribute to Ecosystem Homeostasis -
        1. Growth Factors/Reduction Factors - Ecosystem balance is maintained by the opposing forces of both biotic and abiotic growth and reduction factors.
        2. Environmental Resistance - Environmental resistance is a collective term for those factors which have a negative effect on population growth.

3. Resisting Small Changes Inertia and resilience are characteristics of ecosystems which resist and rapidly recover

C. The Resilience of Ecosystems - In ecosystems, changes in biotic and abiotic factors may perturbate the ecosystem but all ecosystems tend to return to normal over time.
   1. Resilience - The ability of an ecosystem to recover from temporary changes in conditions.

D. Resisting Changes from Human Activities
   1. The resilience of ecosystems minimizes and sometimes negates human impacts. Human impacts can, however, severely disturb an ecosystem and overcome the recuperative capacity of the system.

E. Population Control and Sustainabihty - Environmental resistance helps to keep populations within an ecosystem's carrying capacity.

F. Species Diversity and Stability - Species diversity is a rough measure of the number of species living in a community. Species diversity and stability seem to be positively correlated for most ecosystems, though climate uniformity may also play a major role in ecosystem stability.

II. Natural Succession - Succession is a process whereby one biotic community replaces another.

A. Primary Succession The development of a biotic community where none had existed before is primary succession.
   1. Each successional community (pioneer, intermediate and climax) changes conditions, making them unfavorable for itself and favorable for the next community. The final or longest-lasting stage in succession is the climax community.
   2. Secondary Succession - This takes place when an existing community is disturbed or destroyed, either by natural or human-caused events.
   3. Changes During Succession -
      a. Complexity and efficiency usually increase as ecosystems mature.
      b. Species diversity may peak in either immature or climax stages, depending on the particular ecosystem.
      c. Damage to an ecosystem may be so severe as to essentially prevent recovery by natural succession or restoration.

III. Evolution: Responding to Change - Evolution produces species which are structurally, functionally, and behaviorally adapted to their environment.

A. Natural Selection - Natural selection is a process which favors those individuals who are the most fit or reproductively successful. Natural selection is the driving force or evolution.

B. Genetic Variation - Genetic variation refers to naturally occurring differences in the genetic composition of organisms in a population. This is the raw material of evolution The sources of genetic variation are:
   1. Mutation - Mutations are changes in the DNA .
   2. Gamete production - Genetic material may be transferred from one chromosome to another (crossing over).
   3. Sexual reproduction - Each offspring contains half of each parent's genetic information and thereby new combinations may arise.

C. Speciation - Speciation, the creation of new species, usually occurs as a result of geographic and reproductive isolation of different groups of members of a parent species (allopatric speciation). New species may also form without geographical isolation (sympatric speciation) and is common in plants.

D. Coevolution and Ecosystems Balance - Organisms can evolve in concert with each other. Organisms that are closely associated over long periods of time often have adaptations closely related to the other organism. This results from the combined action of selective pressures each species places on the other though evolutionary time.

IV. Human Impact on Ecosystems - Humans may alter the biosphere by affecting either biotic or abiotic ecosystem components. Tampering with abiotic or biotic factors tends to reduce species diversity and thus simplify ecosystems, which makes them considerably more vulnerable to natural forces. Ecologists refer to the loss of species diversity as *ecosystem simplification*.
   A. Tampering with Abiotic Factors - Human activities produce air and water pollution, which can alter the abiotic environment sufficiently to affect ecosystem balance on local, regional, or global scales. Some pollutants biomagnify or accumulate in increasing amounts as they travel up the food web.
   B. Tampering with Biotic Factors - Destruction or depletion of resources for human use can have serious adverse effects on other species.
      1. Introducing Competitors - Alien species may cause disruption in an ecosystem by out competing native species.
      2. Eliminating or Introducing Predators - Predators play a vital role in regulating ecosystem stability. Their elimination, or their introduction into new habitats, can cause serious ecosystem imbalance.
      3. Introducing Disease Organisms - Introduced pathogens can wipe out native species which are not resistant to the alien disease organism.
   C. Simplifying Ecosystems - Most human-caused ecosystem changes simplify ecosystems by reducing species diversity. Very simple ecosystems, such as monocultures, tend to be highly unstable. Their protection often causes other environmental problems.
   D. Why Study Impacts? Impact models, though imprecise, can serve as frameworks for environmental risk assessment and decision-making. While environmental impact statements are an effort to use modeling to choose preferred options for projects which will affect the environment, they are generally used to find ways to mitigate damage.
      1. Environmental Impact Statement (EIS) - An EIS is required by law on all major projects on federal lands. The EIS summarizes the environmental, social, cultural and archeological impacts of the proposed project.
      2. Least Impact Analysis (sustainability analysis) - This analysis would suggest alternative plans and would be preventive in nature not mitigation.

Suggestions for Presenting the Chapter

- Instructors should strive to provide local examples of primary and secondary succession. Fieldtrips are excellent venues for exploring these concepts.
- Instructors might ask the students to analyze the condition of local predator populations. Does your state promote hunting or trapping of important predators? Stressing the importance of human impact predator populations is a good starting point for discussion.
- An analysis of attitudes toward predators is also an interesting topic for discussion. Why do humans have problems with predators? Why are many predators threatened or endangered with extinction? How can positive attitudes about predators be formed?

- Examine an Environmental Impact Statement (EIS). Environmental Impact Statements are readily available from the federal government for a variety of proposals. For example, the Forest Service is required to produce an EIS every time a new management plan is proposed for a National Forest. These are available by writing your nearest Forest Service office for a free copy.
- An activism project is always an effective and sustainable project for your classes. Involve your classes in a habitat restoration project on campus or in your area. Local environmental groups always need volunteer help to complete their projects.
- Recommended web sites:

Works of Charles Darwin online: http://www.literature.org/authors/darwin-charles/
The Ecological Society of America/Issues in Ecology: http://esa.sdsc.edu/issues.htm
Forest Succession: http://www.anu.edu.au/Forestry/silvinative/daniel/2bground.html
Stephen Jay Gould: http://prelectur.stanford.edu/lecturers/gould/excerpts/
Center for Restoration Ecology: http://gaia1.ies.wisc.edu/cre/

True/False Questions

1. ___ Predators are animals that consume other animals.
2. ___ Ecologists describe growth factors collectively as environmental resistance.
3. ___ The ability of an ecosystem to recover from temporary changes in conditions is called resilience.
4. ___ Species diversity is a measure of the number of extinct species in an ecosystem.
5. ___ Ecosystems can form on barren ground by a process called natural succession.
6. ___ The first community established on barren land is called the climax community.
7. ___ Climax communities are relatively stable ecosystems believed to be the end point of natural succession.
8. ___ Secondary succession occurs after ecological disturbance in an ecosystem.
9. ___ Nutrients are recycled more efficiently in mature ecosystems than in immature or immediate ecosystems.
10. ___ Fossils are preserved bones and imprints of organisms captured in rock.
11. ___ Gametes are small changes in the gene frequency that produce evolutionary change.
12. ___ Fitness is a measure of the availability of new and sometimes favorable genetic combinations.
13. ___ Speciation that occurs as a result of geographic isolation is called sympatric speciation.
14. ___ The process whereby new species arise is called speciation.
15. ___ Killer bees were intentionally introduced to South America in 1956.
16. ___ The study of how ecosystem recovery occurs and how it can be facilitated is called restoration ecology.
17. ___ Organisms that cause diseases are called pathogens.
18. ___ Increasing the number of species in an ecosystems is called ecosystem simplification.
19. ___ All major construction projects on federal land require an analysis of potential impact called an Environmental Impact Statement.
20. ___ Planting many species of crops in the same field is called monoculture.

Fill-in-the-Blank Questions

1. An Environmental _____ Statement assesses the potential environmental damage by projects on federal lands.
2. Disease causing organisms are called _____.
3. _____ is the process where new species arise.
4. _____ speciation occurs when new species arrive as a result of geographic isolation.
5. DNA consists of many segments called _____, each of which plays a specific role in regulating cell structure and function.
6. The germ cells, the sperm and egg, are called _____.
7. _____ reproduction occurs when offspring are produced by union of sperm from males and ova from females.
8. _____ is a measure of the reproductive success of an individual in a population.
9. _____ is the process in which life formed and new life-forms emerge.
10. The first community to become established on barren land is the _____ community.
11. The end-point of natural succession is the _____ community.
12. _____ succession occurs when an ecosystem is disturbed by natural or human causes.
13. Species _____ is a measure of the number of species living in a community.
14. Factors that cause the population to increase are called _____ factors.
15. _____ are animals that consume other organisms.
16. The human body's state of internal constancy is called _____.
17. Weather and the chemical environment are _____ factors effecting the growth of a population.
18. The animals that are eaten by predators are called _____.
19. The abiotic and biotic growth reducing factors are called environmental _____ by ecologists.
20. Natural _____ is a process in which a biotic community forms on a lifeless piece of ground.

Fill-in Key:

1. Impact
2. pathogens
3. speciation
4. allopatric
5. genes
6. gametes

7. sexual
8. fitness
9. evolution
10. pioneer
11. climax
12. secondary
13. diversity
14. growth
15. predators
16. homeostasis
17. abiotic
18. prey
19. resistance
20. Succession

Multiple Choice Questions

1. Animals that consume other organisms are called:
   a. prey
   b. competitors
   * c. predators
   d. parasites
   e. herbivores

2. Which of the following factors would favor the growth of a population?
   a. high reproductive rate
   b. ability to migrate to new habitats
   c. ability to defend territories
   d. ability to adapt to environmental change
   * e. all of the above

3. The term *environmental resistance* is best described as factors which:
   a. stimulate population growth
   b. increase the quality of available habitat
   c. increase the population's reproductive rate
   * d. reduce population size
   e. reduce the effect of parasitism

4. Predation, disease, parasites, competition, and harmful weather conditions all are:
   a. abiotic factors
   b. biotic factors
   c. growth factors
   * d. reduction factors
   e. chemical factors

5. Species diversity measures the number of:
   a. individuals in a community.
   * b. species in a community.
   c. individuals of one species.
   d. genotypes in the population.
   e. mutations in the population.

6. Biological communities that form on barren or lifeless ground are an example of:
   a. natural selection
   * b. primary succession
   c. secondary succession
   d. restoration biology
   e. zootic climax

7. The sequential development of biotic communities after the partial or complete destruction or an existing community is:
   a. natural selection
   b. primary succession
   * c. secondary succession
   d. restoration biology
   e. zootic climax

8. The colonization of a new volcanic island by organisms is a good example of:
   * a. primary succession
   b. secondary succession
   c. restoration biology
   d. climax community
   e. zootic climax

9. _____ are relatively stable ecosystems believed to be the end point of natural succession.
   a. Pioneer communities
   b. Intermediate communities
   * c. Climax communities
   d. Immature ecosystems
   e. Human ecosystems

10. The biological community formed after the eruption of Mt. Saint Helens is a good example of:
   a. climax communities
   b. primary succession
   * c. secondary succession
   d. zootic climax
   e. speciation

11. Lichens growing on rocks are a classic example of:
   a. speciation
   b. natural selection
   c. competition
   * d. primary succession
   e. secondary succession

12. Environmental resistance is usually the **lowest** during this period of natural succession:
   * a. pioneer community
   b. intermediate community
   c. climax community
   d. the peak of the intermediate community
   e. the peak of the climax community

13. Which of the following factors would be a characteristic of the immature ecosystem?
   a. web-like food chains
   b. low net productivity
   c. high species diversity
   d. good nutrient conservation
   * e. low stability

14. Which of the following factors would be a characteristic of the mature ecosystem?
   * a. weblike food chains
   b. high net productivity
   c. low species diversity
   d. poor nutrient conservation
   e. low stability

15. The structural, functional and behavioral characteristics of an organism that give it a reproductive advantage are called:
   a. traits
   b. gametes
   * c. adaptations
   d. mutations
   e. genotypes

16. The reproductive cells are called:
   a. chromosomes
   b. genes
   * c. gametes
   d. gonads
   e. mutants

17. Genetic variation arising from DNA transfer between chromosomes during the production of the reproductive cells is called:
   a. independent assortment
   b. mutation
   * c. crossing over
   d. natural selection
   e. sexual reproduction

18. The process of forming new species is called:
   a. sexual reproduction
   b. natural selection
   c. crossing over
   * d. speciation
   e. adaptation

19. Formation of new species in different geographical areas that are reproductively isolated is called _____ speciation.
   * a. allopatric
   b. sympatric
   c. reproductive
   d. geographic
   e. adaptive

20. Which of the following are abiotic factors?
   a. competition
   b. predation
   c. parasitism
   * d. weather
   e. population size

21. The scientific study of how ecosystem recovery occurs and how it can be facilitated is called:
    a. population ecology
    b. theoretical ecology
*   c. restoration ecology
    d. environmental studies
    e. human ecology

22. Disease causing microorganisms are called:
    a. mutagens
    b. carcinogens
*   c. pathogens
    d. plankton
    e. detritus

23. The loss of species diversity is called:
    a. reclamation
    b. turn over
    c. environmental resistance
*   d. ecosystem simplification
    e. monoculture

24. Industrial agriculture often uses only one species of crop in a field. This agricultural practice is called:
    a. horticulture
    b. silvaculture
    c. aquaculture
    d. hydroponics
*   e. monoculture

25. Which of the following responses is **not** one of the problems with the practice of monoculture?
    a. low genetic diversity
    b. susceptibility to insects and pathogens
    c. overuse of pesticides
    d. vulnerable to drought and wind
*   e. high yields

26. A state of dynamic equilibrium in a system is called:
*   a.. homeostasis
    b. dynamic balance
    c. carrying capacity
    d. set point
    e. static balance

27. Which of the following is **not** a biotic factor:
*   a. availability of water
    b. competition
    c. predation
    d. parasitism
    e. disease

28. The prairie vole is a small:
    a. bird
    b. weasel
*   c. rodent
    d. lizard
    e. snake

29. The ability of an ecosystem to recover from temporary changes is called:
    a. reclamation
    b. diversity
*   c. resilience
    d. negative feedback
    e. sustainability

30. Which of the following factors have the potential for controlling population size?
    a. predation
    b. competition
    c. parasitism
    d. disease
*   e. all of the above

31. The population size that can be supported by a particular ecosystem is called the:
    a. set point
    b. climax
*   c. carrying capacity
    d. biotic factor
    e. abiotic factor

32. The highest biological diversity would be expected to occur in:
    a. polar regions
    b. temperate regions
    c. taiga
*   d. low latitudes
    e. marine ecosystems

33. In mature ecosystems, one would expect to find:
    a. low diversity and high stability
    b. low diversity and low stability
*   c. high diversity and high stability
    d. high diversity and low stability
    e. linear food chains and low stability

34. In mature ecosystems, one would expect to find low:
    a. species diversity
    b. stability
    c. nutrient conservation
*   d. net productivity
    e. all of the above

35. Avalanche areas in the rocky mountains are first colonized by:
    a. fir
    b. spruce
    c. willows
*   d. grasses
    e. larch

36. Preserved bones and imprints of organisms captured in rock are called:
    a. strata
    b. detritus
*   c. fossils
    d. artifacts
    e. seston

37. The theory of evolution by natural selection was proposed by:
    a. Watson and Crick
*   b. Darwin and Wallace
    c. William Tucker
    d. Robert Gilman
    e. Ed Garbisch

38. Killer bees were brought from Africa to South America and reached the United States in the mid _____.
    a. 1950's
    b. 1960's
    c. 1970's
*   d. 1980's
    e. 1990's

39. This organism was accidentally introduced into North America by tankers arriving from Europe:
    a. water hyacinth
    b. killer bees
    c. grass carp
*   d. zebra mussel
    e. red deer

40. This species has been introduced from the southeastern United States into many subtropical regions because it eats insect larvae and thus helps to control malaria:
    a. grass carp
*   b. mosquito fish
    c. emerald shiner
    d. tench
    e. channel catfish

8

Human Ecology: Our Changing Relationship with the Environment

Chapter Outline

Human Biological Evolution

Human Cultural Evolution
    Hunting and Gathering Societies
    Agricultural Societies
    Industrial Society
    Advanced Industrial Age

The Population, Resources and Pollution Model

The Sustainable Society: The Next Step

Key Terms

dryopithecines	australopithecines	bipedal
hunting and gathering	agricultural societies	slash-and-burn
agriculture	industrial society	Industrial Revolution
agroforestry	advanced industrial society	biodegradable
nonbiodegradable	cross-media contamination	negative feedback loop
positive feedback loop	Sustainable Revolution	

Objectives

1. Discuss the primary adaptations that have evolved during human evolution that give our species a distinct competitive advantage over many other species.
2. List the major societies that have evolved during human evolution and compare their characteristics.

3. Discuss the technique of agroforestry and its advantages for sustainable use of tropical forest areas.
4. Discuss the impact of Industrial Societies on the environment.
5. Define the following terms: "biodegradable", "nonbiodegradable", "cross-media contamination", "negative feedback loop", and "positive feedback loop".
6. Discuss the importance of the "Sustainable Revolution" for creating a sustainable future.

Lecture Outline

I.   Human Biological Evolution - Humans and their fossil relatives are referred to as hominids The upright stance and large brain size of hominids facilitated the evolution of certain features, such as language and manual dexterity, which have dramatically expanded our capacity to alter nature.

II. Human Cultural Evolution - All human societies have impacted the environment, the extent of the impact depending on the society's population size, resource demands, and choice of resources for support.
   A. Hunting and Gathering Societies - For most of our history, humans have lived in hunting and gathering societies. Members of these societies apparently had a deep reverence for, and knowledge of, their environment. Though not always environmentally benign, these societies, due to their low population size and primitive technology, were generally sustainable.
   B. Agricultural Societies - The Agricultural Revolution occurred around 8000 b.c., beginning probably in Southeast Asia. Domestication of plants and animals allowed populations to increase, cities to develop, and humans to change their relationship with nature from one of cooperation and respect to one of domination and exploitation.
   C. Industrial Societies - These societies rose, to prominence out of the Industrial Revolution of the 18th and 19th Centuries. Increasing mechanization, new technology, and improved sanitation all set the stage for rapid population growth and environmental deterioration, a result of physical and spiritual disconnection from the earth.
   D. Advanced industrial societies arose following World War II and are characterized by high levels of production, consumption, heavy reliance on synthetics and nonrenewables, and increasing per capita energy demand. The product of extreme industrialization has been societies physically and spiritually alienated from the land, increasingly environmentally destructive, and living unsustainably.

III. The Population, Resources, and Pollution Model - The PRP Model shows that human populations acquire and use resources and in so doing negatively impact the air, water, and soil.
   A. Human-generated pollutants can be biodegradable or nonbiodegradable; when they cross boundaries between environmental media, they constitute *cross-media contamination.*
   B. Both negative and, to a lesser extent, positive feedback loops are activated as a result of human manipulation of the environment.

IV. The Sustainable Society: The Next Step - The next step in cultural evolution is the creation of a sustainable society, the Sustainable Revolution. Many changes are already under way and are starting to restructure human systems for sustainability.

Suggestions for Presenting the Chapter

- Viewing a video on human anthropology/cultures and is a stimulating way to supplement the text material.
- Instructors should emphasize and explore the change in value systems associated with the cultural transitions from hunting/gathering to agricultural, industrial and advanced/post-industrial societies.
- The Population, Resources, and Pollution model can be used in a class exercise. Groups can be assigned to each component of the model and can identify important activities involved in their component and how they affect the other components of the model.
- Students can be assigned the task to identify biodegradable and nonbiodegradable products used at home or at their educational institution. A discussion of alternatives to nonbiodegradable products or ways to prevent use of unnecessary products is highly recommended.
- Recommended web sites:

Human Origins and Evolution in Africa: http://www.indiana.edu/~origins/
Paleoanthropology Links: http://www.talkorigins.org/faqs/homs/links.html
International Center for Research in Agroforestry: http://www.icraf.cgiar.org/
Agroforestry for Farms and Ranches: http://www.nhq.nrcs.usda.gov/BCS/forest/tnote1.html
Smithsonian Institution Human Origins Program:http://www.mnh.si.edu/anthro/humanorigins

True/False Questions

1. ___ Dryopithecines have rise to modern apes (chimpanzees, gibbons, and gorillas).
2. ___ *Homo sapiens* emerged about 100,000 years ago.
3. ___ Walking by using two limbs is called bipedal locomotion.
4. ___ Hunting and gathering people had a nomadic lifestyle.
5. ___ Slash-and-burn agriculture is also called swidden agriculture.
6. ___ The plow came into use about 20,000 years B.C..
7. ___ The first area to develop seed crop agriculture was in Australia.
8. ___ Agroforestry is a method used to harvest trees for food production.
9. ___ Industrial societies emerged in England in the 1700s.
10. ___ Individuals in hunting and gathering societies require about 2000-5000 kilocalories per day.
11. ___ DDT is a nutrient added to animal feed to increase growth.
12. ___ Biodegradable materials are broken down by bacteria and other organisms.
13. ___ Acid rain is an example of cross-media contamination.
14. ___ A kilocalorie is a measure of length.

15.___ Per capita energy use in the United States and Canada averages about 2500 kilocalories per day.
16.___ John Deere tractors were important in the development of agricultural societies.
17.___ Fossil fuels were very important in hunting and gathering societies.
18.___ Slash-and-burn agriculture is used in tropical rain forests to create cropland.
19.___ Australopithecines emerged about 3.5 million years ago.
20.___ Mass production and modern technology are part of industrial societies.

True/False Key:

1. T 2. F 3. T 4. 5. T 6. F 7. F 8. F 9. T 10. T 11. F 12. T 13. T 14. F 15. F 16. F 17. F 18. T 19. T 20. T

Fill-in-the-Blank Questions

1. _____ gave rise to the modern apes.
2. _____ arose about 3.5 million years ago.
3. Walking with two legs is called _____ locomotion.
4. During 99% of the time the human species have been on earth we were _____ and _____.
5. Hunters and gatherers were _____ and roamed the land to make a living.
6. Agricultural societies emerged between _____ and 6000 B.C..
7. The roots of agriculture are generally traced to the rain forests of Southeast _____.
8. The plow came into use about _____ B.C.
9. Agroforestry is a system that combines agriculture with ____ crops or other forest plants.
10. In _____ agriculture, people clear small plots or rain forest to plant food crops or perennial tree crops.
11. The _____ revolution started in England in the 1700s and the United States in the 1800s.
12. One kilocalorie is equal to _____ calories.
13. Early agricultural societies required _____ the amount of energy per person than hunting and gathering societies.
14. Materials that can be broken down by bacteria and other organisms are said to be _____.
15. Air pollution that washes from the sky into a lake is called _____ contamination.
16. The shift toward the use of _____ materials is a characteristic of advanced industrial societies.
17. DDT is a persistent _____.
18. Per capita energy use in the United States and Canada is about _____ kilocalories per day.
19. Scientists classify pollutants by the _____ they contaminate, for example, air, water and land.
20. The Exxon Valdez oil spill off the coast of Alaska occurred in _____.

Fill-in Key:

1. dryopithecines
2. australopithecines
3. bipedal
4. hunters, gatherers
5. nomadic
6. 10,000
7. Asia
8. 3000
9. tree
10. swidden
11. Industrial
12. 1000
13. twice
14. biodegradable
15. cross-media
16. synthetics
17. pesticide
18. 250,000
19. medium
20. 1989

Multiple Choice Questions

1. Humans evolved directly from this mammalian group:
   a. marsupials
 * b. primates
   c. rodents
   d. bovids
   e. canids

2. This humanlike group emerged about 3.5 million years ago and are one of our ancestors:
 * a. australopithecines
   b. dryopithecines
   c. gibbons
   d. orangutans
   e. lemurs

3. This adaptation allowed our ancestors to use tools and weapons and to manipulate objects:
   a. quadripedal locomotion
 * b. bipedal locomotion
   c. long arms
   d. prehensile tail
   e. carnassial apparatus

4. *Homo sapiens*, the species to which humans belong, emerged about _____ years ago:
   a. 100,000
   b. 200,000
   c. 300,000
 * d. 400,000
   e. 500,000

5. This stage of human cultural development is characterized by small nomadic groups that survive by scavenging, hunting and collecting fruits, seeds, and berries:
   a. agricultural societies
   * b. hunting and gathering societies
   c. industrial societies
   d. sustainable societies
   e. postindustrial societies

6. This stage of cultural development is characterized by planting crops and use of domesticated animals:
   a. hunting and gathering societies
   * b. agricultural societies
   c. industrial societies
   d. advanced industrial societies
   e. sustainable societies

7. Farmers clear small sections of the forest and often burn the area before planting crops in this agricultural technique:
   * a. swidden agriculture
   b. monoculture
   c. biointensive organic farming
   d. green revolution
   e. conservation tillage

8. The first plows came into use about:
   a. 500 A.D.
   b. 1000 B.C.
   c. 2000 B.C.
   * d. 3000 B.C.
   e. 4000 B.C.

9. A sustainable agricultural technique that combine agriculture with tree crops or other forest plants is called:
   a. biointensive organic farming
   b. conservation tillage
   * c. agroforestry
   d. ridge tillage
   e. industrial agriculture

10. Industrial societies emerged in England in the:
    a. 1500's
    b. 1600's
    * c. 1700's
    d. 1800's
    e. 1900's

11. The Industrial Revolution was made possible by new technologies but also by an abundant source of:
    a. labor
    b. machines
    * c. fuel
    d. capital
    e. available goods

12. Which countries are mentioned in the text as users of agroforestry?
    * a. Peru and Indonesia
    b. Canada and New Zealand
    c. United States and Canada
    d. France and Germany
    e. Denmark and Sweden

13 Who said people must become "emancipated from the bonds of nature?"
    a. Charles Darwin
    * b. John Locke
    c. Cyrus McCormick
    d. Jethro Wood
    e. Dale Carnegie

14. A kilocalorie is a measure of:
    a. length
    b. mass
    c. volume
    d. density
    * e. energy

15. A shift toward the use of synthetic materials occurred in which of the states of cultural development?
    a. hunting and gathering societies
    b. agricultural societies
    c. industrial societies
    * d. advanced industrial societies
    e. sustainable societies

16. A kilocalorie is equal to _____ calories.
    a. 10
    b. 100
    * c. 1000
    d. 10,000
    e. 100,000

17. Hunter-gatherers used about _____ kilocalories per person per day.
* a. 2000 to 5000
   b. 8000 to 10,000
   c. 20,000 to 30,000
   d. 50,000 to 60,000
   e. 120,000 to 150,000

18. Modern industrial society uses about _____ kilocalories per person per day.
   a. 2000 to 5000
   b. 10,000
   c. 20,000
   d. 60,000
* e. 120,000

19. Pollutants that can be broken down by bacteria and other organisms in the environment are:
   a. sustainable
   b. nonbiodegradable
* c. biodegradable
   d. recyclable
   e. biosolids

20. Air pollution that washes form the sky and is deposited in lakes and forests causes:
   a. air pollution
* b. cross-media contamination
   c. negative feedback
   d. positive feedback
   e. nonbiodegradable pollution

21. Which of the following is **not** a characteristic of hunting and gathering societies?
   a. nomadism
   b. manipulation of tools and weapons
   c. used fire
* d. farming
   e. knowledgeable about the environment

22. Which of the following is a characteristic of agricultural societies?
   a. farming
   b. use of domestic animals
   c. subsistence or urban based
   d. highly exploitive of resources
* e. all of the above

23. Which of the following is **not** a characteristic of industrial societies?
   a. mass production
   b. highly exploitive of resources
   c. environmental impacts are serious
   d. relies on new technologies
* e. nomadic

24. Human populations were hunter-gatherers about _____ % of the time we have spent on earth.
   a. 25
   b. 50
   c. 75
   d. 90
* e. 99

25. Anthropologists believe that agricultural societies merged about _____ years ago.
   a. 1000
   b. 5000
* c. 10,000
   d. 20,000
   e. 50,000

26. What percent of the world's existing rainforest is cut down each year for population growth?
   a. 10
   b. 15
* c. 25
   d. 35
   e. 50

27. According to the Population, Resources, and Pollution Model, soil erosion would form a/an:
   a. positive feedback loop
* b. negative feedback loop
   c. closed loop system
   d. open loop system
   e. reduced environmental resistance

28. According to the Population, Resources, and Pollution Model, efficient harvest of food and fiber would form a/an:
* a. positive feedback loop
   b. negative feedback loop
   c. closed loop system
   d. open loop system
   e. increased environmental resistance

29. The social system where humans become subject to ulcers, heart disease and mental illness is:
    a. hunting and gathering societies
    b. agricultural societies
    * c. industrial societies
    d. tribal societies
    e. matriarchal societies

30. The roots of modern agriculture have been traced back to the rain forests of:
    a. Central America
    b. South America
    c. Brazil
    d. Australia
    * e. Southeast Asia

31. The Babylonian Empire was once covered with productive forests and grasslands and collapsed as a result of environmental damage. This empire once occupied the land that is now part of the countries of:
    a. Israel and Jordan
    b. Saudi Arabia and Sudan
    c. Morocco and Libya
    * d. Iraq and Iran
    e. Turkey and Greece

32. Which of the following is **not** a characteristic of advanced industrial societies?
    a. high per capita energy consumption
    b. use of fossil fuels
    c. use of synthetics
    d. dominion of nature
    * e. sustainable human systems

33. In a speech before the United Nations this person called our home "Spaceship Earth"?
    a. Theodore Roosevelt
    b. John Kennedy
    c. Sir Frederick Hoyle
    * d. Adlai Stevenson
    e. Al Gore

34. Slash-and-burn agriculture is a type of:
    a. green revolution technology
    b. ridge tillage
    c. conservation tillage
    * d. agroforestry
    e. silvaculture

35. The Mayans of Mexico were probably the first people in the western hemisphere to:
    a. become hunter-gatherers.
    b. use plows.
    * c. grow seed crops.
    d. abuse agricultural land.
    e. convert cropland to rangeland.

36. Centers of early civilization seem to be near:
    a. regions of low diversity.
    b. desert areas.
    c. the equator.
    * d. centers of origin for seed crops.
    e. center of origin for industrialization

37. Farmers clear forests and plant teak trees whose wood is exported to many developed nations. Among the teak trees, rice and corn is planted; the country practicing this agroforestry technique is:
    a. Peru
    b. Brazil
    * c. Java
    d. Philippines
    e. Cuba

38. Which of the following are sustainable advantages of agroforestry?
    a. protection of virgin forest
    b. sustainable food supply
    c. sustainable fuel supply
    d. reduces soil erosion
    * e. all of the above

39. Using the Population, Resources, and Pollution model, if population increases:
    a. pollution decreases
    b. resource use decreases
    * c. resource acquisition increases
    d. population size will continue to increase
    e. population size will become stable

40. Cross-media contamination occurs when pollutants move from :
    a. surface water to ground water
    b. car exhaust to outside air
    c. landfill water moves to ground water
    d. industrial water pollutants move to streams
    * e. air pollution becomes acid rain

9

Population: Measuring Growth and Its Impact

Chapter Outline

The Growing Human Population
Why Has the Human Population Grown So Large?
Expanding the Earth's Carrying Capacity
What is the Earth's Carrying Capacity for Humans?
Too Many People, Reproducing Too Quickly

Understanding Populations and Population Growth
Measuring Population Growth
Total Fertility rate and Replacement -Level Fertility
Migration
Population Histograms
Exponential Growth

The Future of World Population
Why should we worry about population growth in the developing world?
A World of Possibilities

Key Terms

population growth	carrying capacity	infant mortality
overpopulation	growth rate	crude birth rate
crude death rate	doubling time	total fertility rate
replacement-level fertility	zero population growth	migration
immigration	emigration	internal migration
exponential growth	population histogram	sigmoidal

Objectives

1. Discuss why the human population has grown so large.
2. Define "carrying capacity".

3. Discuss the impact of overpopulation on our environmental problems.
4. Define the following terms: "growth rate", "crude birth rate", "crude death rate", and "doubling time".
5. Compare the population growth rates of developing and developed nations.
6. Define "total fertility rate" and "replacement-level fertility".
7. Discuss why population size in the United States continues to grow even though the replacement-level fertility rate has been reached or exceeded.
8. Define "migration", "immigration", "emigration" and "net migration".
9. Discuss the pros and cons of immigration in the United States.
10. Describe a population histogram and explain its use in understanding the demography of a population.
11. Draw an exponential growth curve and a sigmoid curve.

Lecture Outline

I. The Growing Human Population - Currently, the world population is 5.84 billion; 100 million people are added yearly.
   A. Why Has the Human Population Grown So Large?
      1. The Survival Boom - The phenomenal upsurge in population growth in modern times is due to two factors: slightly increased reproduction rates and dramatically decreased death rates, which have contributed to a higher average life expectancy.
   B. Expanding the Earth's Carrying Capacity - The maximum population an ecosystem can support sustainably is termed the *carrying capacity*.
      2. Advances in tools, agriculture, medicine, and technology have enabled humans to increase the earth's carrying capacity.
      3. While necessary to provide for an expanding human population, such increases invariably result in increased depletion and pollution.
   C. What is the Earth's Carrying Capacity for Humans? - Accurate estimates of the earth's carrying capacity for humans are difficult due to the variables of affluence and technology. Many people that the human population already exceeds the Earth's long-term carrying capacity, creating an unsustainable condition.
   D. Too Many People, Reproducing Too Quickly - Overpopulation is at the root of virtually all environmental problems, including pollution and resource depletion, and many social and economic problems. The results are:
      1. Urban despair from overpopulation.
      2. Rural despair caused by overpopulation.
      3. All of the social, economic and environmental problems are exacerbated by rapid growth. Efforts to solve these problems are hampered by rapid population growth.

II. Understanding Populations and Population Growth - Demography is the study of population statistics and characteristics.
   A. Measuring Population Growth
      1. Growth Rate - Growth rate (%) = (Crude Birth Rate - Crude Death Rate) x 100

2. Birth Rates - Birth rates for any population are influenced by age of marriage, educational level, contraceptive use, female employment, and couples' desires, beliefs and values.
3. Death Rates - Rapidly falling death rates due to improved living conditions and medical treatment have caused an increase in growth rate in the past century.
4. Doubling Time - The formula for doubling time is: $DT = 70/GR(\%)$. At the current growth rate of 1.8%, developing nations will double in about 33 years. Growth rates in developing countries on average far exceed those of developed nations.

B. The Total Fertility Rate and Replacement -Level Fertility
1. The Total Fertility Rate (TFR) is the average number of children a woman will bear.
2. Replacement-level fertility is reproduction at exactly that level necessary for couples to replace themselves.
3. Due to immigration and momentum from the lag effect, most populations will not reach zero population growth immediately upon reaching replacement reproduction.

C. Migration - Migration is the number of individual moving in a population. Migration has two forms, immigration (movement into a country or system) and emigration (movement out of a country or system).
1. Net migration is immigration minus emigration.
2. Zero population growth for any population can only be reached when growth rates and net migration rates are zero.
3. Intranational or internal migration has had profound demographic, as well as economic and environmental effects in the U.S.

D. Population Histograms - Histograms are graphic representations of the age structure and gender distribution of a population. Shifts in population structure can signal dramatic changes in the economic, employment, infrastructure and healthcare needs of a population

E. Exponential Growth
1. Growth at a fixed yearly rate is exponential.
2. When graphed, exponential growth can be seen to start slowly, then round a bend to enter a phase of increasingly rapid growth; if unchecked, an overshoot may occur. Resource recovery and pollution assimilation cannot keep up with the rapid growth in human population.

III. The Future of World Population
A. Estimates of maximum human population size range from 8 to 15 billion; it may then level off, decrease slowly, or drop dramatically, depending on many variables.
1. No finite system can accommodate infinite population growth.
B. Why should we worry about population growth in developing nations?
1. Illegal immigration into the United States.
2. Increased political strife in the Middle East and elsewhere.
3. Growing international population increases the production of food for export. This places additional stress on our already unsustainable agricultural system.
4. Expanding world population results in destruction of habitat and loss of valuable species.
5. Population growth exacerbates global warming.
6. The cycle of poverty is revisited in developing nations.

Suggestions for Presenting the Chapter

- The instructor should stress that many of the world's problems have their root in population size. Overpopulation affects every part of our lives.
- A speaker on family planning will provide interesting topics for class discussion.
- The instructor should emphasize that although developing nations are not growing rapidly they still have population concerns because of the per capita impact of individuals in these countries.
- Computer software for modeling population dynamics is available. Exposing students to these programs are an effective tool for teaching concepts in this chapter.
- Recommended web sites:

  Population Ecology: http://www.gypsymoth.ento.vt.edu/~sharov/popechome/welcome.html
                      http://ecology.about.com/science/ecology/cs/populationecology/index.htm
  Carrying Capacity - Area-Based Indicators of Sustainability: http://dieoff.org/page110.htm
  Population Dynamics Models: http://rulbii.leidenuniv.nl/wwwkim/popdyn.html
  Centre for Population Studies: http://www.lshtm.ac.uk/eps/cps/cpslinks.htm

True/False Questions

1. ___ The average number of years a person lives is called the life expectancy.
2. ___ The number of organisms that an area can support is called the carrying capacity.
3. ___ The human population is estimated to use about 1% of the earth's terrestrial primary productivity.
4. ___ Overpopulation contributes significantly to the troubles of cities making many of them unsustainable.
5. ___ Inner city syndrome has been suggested to occur because of low population densities in the inner city.
6. ___ The growth rate of a population is equivalent to the crude birth rate.
7. ___ The crude death rate is the number of deaths per 1000 people.
8. ___ The crude birth rate is the number of births per 10,000 people.
9. ___ Thailand adopted a national population policy that has been successful in reducing the countries growth rate.
10. ___ The doubling time is the amount of time to reduce population size by half.
11. ___ The total fertility rate is the number of children women in a population are expected to have during their lifetimes.
12. ___ Immigration refers to the movement of people out of a country.
13. ___ The movement of people within a country from one region to another is called internal migration.
14. ___ ZPG stand for zero population growth.
15. ___ A population histogram displays the birth rate of a population.
16. ___ Doubling time is calculated by dividing 70 by the growth rate.
17. ___ Overpopulation results when a population exceeds the Earth's carrying capacity.
18. ___ Technological advances have increased environmental resistance decreasing the Earth's carrying capacity for humans.

19. ___ Birth rates depend on the age men and women get married.
20. ___ Overpopulation is not a root cause of the environmental crisis.

True/False Key:

1. T 2. T 3. F 4. T 5. F 6. F 7. T 8. F 9. T 10. F 11. T 12. F 13. T 14. T
15. F 16. T 17. T 18. F 19. T 20. F

Fill-in-the-Blank Questions

1. _____ growth occurs when the population increases by a fixed percentage each year.
2. A population _____ is a bar graph that displays the age and sex composition of a population.
3. When the _____ rate equals the death rate the population will cease to grow.
4. ZPG stands for _____ population growth.
5. Total _____ rate is the number of children women in a population are expected to have in their lifetime.
6. _____ refers to the movement of people into a country.
7. _____ refers to the movement of people out of the country.
8. _____ migration refers to the movement of people within a country from one region to another region.
9. The United States, Japan, Australia and Russia are _____ countries with a strong economic base.
10. _____ time is the time it takes a population to double in size.
11. The number of births per 1000 people in a population is the _____ birth rate.
12. The growth rate is equal to the difference between the crude birth rate and the crude _____ rate.
13. The _____ capacity is the number of organisms a particular area can support.
14. Social problems as a result of overpopulation in cities has been called inner city _____.
15. The world's fastest growing countries in decreasing order are: _____, Latin American, and Asia.
16. Populations can continue to grow after they have reached replacement-level fertility due to the _____ effect.
17. TPR refers to the total _____ fertility.
18. In 1971 _____ adopted a national population policy that was successful in reducing the population's growth rate.
19. S-shaped growth curves are also called _____ curves.
20. In _____ Congress passed the Immigration Act.

Fill-in Key:

1. exponential

78

2. histogram
3. birth
4. zero
5. fertility
6. immigration
7. emigration
8. internal
9. developed
10. doubling
11. crude
12. death
13. carrying
14. syndrome
15. Africa
16. lag
17. replacement
18. Thailand
19. sigmoidal
20. 1990

Multiple Choice Questions

1. What percent of the current population growth is occurring in the developing countries?
   a. 20
   b. 40
   c. 60
   d. 80
   * e. 90

2. In 1997 the world population reached ____ billion people.
   a. 4.5
   * b. 5.8
   c. 6.5
   d. 7.2
   e. 8.5

3. Which of the following explain why the human population has continued to grow at exponential rates?
   a. high birth rates
   b. increases in the food supply
   c. better medicine
   d. sanitation
   * e. all of the above

4. The human population reached the size of one billion by the year:
   a. 1750
   * b. 1850
   c. 1900
   d. 1935
   e. 1975

5. The world population is predicted to reach eight billion people by the year ____.
   a. 1999
   b. 2005
   * c. 2017
   d. 2050
   e. 2100

6. The increase in the life expectancy at birth is a primarily the result of:
   a. people living longer
   * b. low infant mortality
   c. better medical care
   d. higher reproductive rates
   e. reduced mortality during senescence

79

7. Which of the following events have reduced the environmental resistance on population growth?
   a. Development of tools and weapons
   b. The Agricultural Revolution
   c. The Industrial Revolution
   d. The opening of the New World for colonization
   * e. all of the above

8. Overpopulation is at the root of virtually all environmental problems including:
   a. pollution
   b. resource depletion
   c. social problems
   d. economic problems
   * e. all of the above

9. The harmful sociological effects associated with overcrowding in urban areas are called:
   a. poverty
   b. urban sprawl
   * c. inner city syndrome
   d. cultural transition
   e. societal transition

10. In 1990 ___% of the world's population lived in cities.
    a. 15
    b. 30
    * c. 45
    d. 60
    e. 75

11. The world's fastest growing area is:
    * a. Africa
    b. Latin America
    c. Asia
    d. India
    e. South America

12. The number of births per 1000 people in a population is the:
    a. growth rate
    b. death rate
    * c. birth rate
    d. doubling time
    e. instantaneous growth rate

13. The birth rate of a population depends on which of the following factors?
    a. the age of marriage
    b. educational level of the parents
    c. whether the woman works after marriage
    d. use of contraceptives
    * e. all of the above

14. If a developing country has a growth rate of 2.8% and doubles its population size in 24 years, how long would it take to double population size with a growth rate of 5.6%?
    a. 6 years
    * b. 12 years
    c. 24 years
    d. 36 years
    e. 48 years

15. If a developing country has a doubling time of 45 years and a growth rate of 1.5%, what would the growth rate of a country be that doubles every 22.5 years?
    a. 0.75%
    b. 1.5%
    c. 2.25%
    * d. 3.0%
    e. 4.5%

16. Which of the following is **not** a characteristic of developing countries?
    a. low standard of living
    b. low doubling time
    c. high infant mortality
    * d. high industrialization
    e. low energy use per capita

17. Which of the following is **not** a reason for the success of Thailand's national population policy?
    a. Buddhist religion
    b. distribution of contraceptives
    c. cultural openness
    d. financial support of government
    * e. mandatory contraceptive use

18. Doubling time is calculated by dividing ____ by the growth rate.
    a. 15
    b. 35
*   c. 70
    d. 140
    e. 280

19. This is the most popular form of contraception in Thailand:
    a. condoms
    b. IUDs
    c. diaphragms
    d. birth control pills
*   e. sterilization

20. The growth rate of the United States population is approximately ___ %.
*   a. 0.5
    b. 1.0
    c. 1.5
    d. 2.0
    e. 2.5

21. Which of the following countries are actually experiencing negative growth rates?
    a. Japan and China
    b. Cuba and Iran
*   c. Germany and Hungary
    d. Canada and Australia
    e. Brazil and Argentina

22. Which of the following is **not** a characteristic of a developed country?
    a. high standard of living
*   b. low per capita food intake
    c. low growth rate
    d. low infant mortality
    e. high urban population

23. The number of children women are expected to have during their lifetimes is the:
    a. crude growth rate
    b. replacement-level fertility
*   c. total fertility rate
    d. fecundity
    e. clutch size

24. The total fertility rate at which couples produce exactly the number of children needed to replace themselves is the:
    a. crude birth rate
*   b. replacement-level fertility
    c. stable age distribution
    d. clutch size
    e. fecundity

25. The replacement-level fertility rate in the United States and other developed countries is ____ children.
    a. 1.0
    b. 2.0
*   c. 2.1
    d. 2.3
    e. 2.5

26. A population stops growing when the birth rate equals the death rate and when net migration is zero. This state is called:
    a. stable age distribution
*   b. zero population growth
    c. maximum sustainable yield
    d. replacement-level fertility
    e. total fertility rate

27. The movement of people within a country from one region to another is:
*   a. internal migration
    b. emigration
    c. immigration
    d. net migration
    e. urban sprawl

28. Human populations are growing:
    a. in a linear fashion.
    b. arithmetically.
*   c. exponentially
    d. toward a stable age distribution.
    e. toward zero population growth.

29. Which of the following responses best explains why the U.S. population is growing even though the total fertility rate is below the replacement-level fertility?
    a. lag effect
    b. immigration is greater than emigration
    c. reduced infant mortality
*   d. a and b only
    e. b and c only

81

30. The movement of people into and out of a population is:
* a. migration
  b. immigration
  c. emigration
  d. mortality
  e. internal migration

31. Movement of people into a country is:
  a. migration
* b. immigration
  c. emigration
  d. mortality
  e. internal migration

32. Movement of people out of a country is:
  a. migration
  b. immigration
* c. emigration
  d. mortality
  e. internal migration

33. _____ accounts for 30% to 40% of the annual growth of the U.S. population each year.
  a. emigration
* b. immigration
  c. births
  d. lag effect
  e. internal migration

34. The 1990 Immigration Act:
  a. Increased the number of illegal aliens.
  b. Prohibited illegal aliens from entering.
* c. Increased legal immigration by 35%.
  d. Decreased legal immigration by 35%.
  e. Increased illegal immigration by 35%.

35. An estimated _____ illegal aliens enter the U.S. annually.
  a. 100,000
  b. 200,000
  c. 300,000
  d. 400,000
* e. 500,000

36. The most likely projections calls for an increase of over ____ million Americans by the year 2050.
  a. 32
  b. 65
* c. 130
  d. 165
  e. 195

37. Which of the following is a potential environmental effect of overpopulation?
  a. habitat destruction
  b. loss of species
  c. global climate change
  d. poverty
* e. all of the above

38. Overshoot occurs when population growth exceeds the _____ of an ecosystem.
  a. available space
* b. carrying capacity
  c. environmental resistance
  d. stability
  e. primary productivity

39. The human population taps about ___% of the earth's terrestrial primary productivity.
  a. 10
  b. 20
  c. 30
* d. 40
  e. 50

40. The condition that results when the human population exceeds the earth's carrying capacity is:
  a. exponential growth
  b. lag effect
* c. overpopulation
  d. inner city syndrome
  e. urban sprawl

10

Stabilizing the Human Population: Strategies for Sustainability

Chapter Outline

Achieving a Sustainable Human Population: The Challenges

Stabilizing the Human Population: Some Strategies
    Economic Development and the Demographic Transition
    Family Planning and Population Stabilization
    Small-Scale, Sustainable Economic Development
    Sustainable Populations in the Developed Countries
    Creating Sustainable Populations in the Developing Countries

Overcoming Barriers
    Psychological and Cultural Barriers
    Educational Barriers
    Religious and Cultural Barriers

Ethics and Population Stabilization

Status Report

Key Terms

contraceptive measures
birth control
environmental impact
population control

demographic transition
abstinence
collective rights

family planning
induced abortion
ethics

Objectives

1. Discuss why demographic transition may not work to reduce population growth in some developing countries.
2. Define "family planning" and list the different methods of birth control.
3. Compare extended voluntary family planning to forced family planning.
4. Define "urban sprawl" and discuss how it relates to growth management.
5. Discuss why populations of the developed nations have a dramatic impact on the environment.
6. List the population control strategies for developed countries.
7. List the population control strategies for developing countries.
8. Discuss how psychological, economic, educational, religious, and cultural factors can influence population growth.
9. Contrast the view of individual rights with the idea of collective rights as they pertain to reproductive freedom.

Lecture Outline

I. Achieving a Sustainable Human Population: The Challenges
   A. Finding acceptable means of slowing the growth of the human population in order to create a stable population size.
   B. Find ways to humanely reduce the size of the human population.
   C. Stopping population growth, reducing population size and developing sustainable economic plans can help break the viscous cycle of poverty and environmental destruction.

II. Stabilizing the Human Population: Some Strategies - Stabilizing the human population will require attacking the root causes of rampant population growth: poverty, lack of education, the inequality of women, and poor health care.
   A. Economic Development and the Demographic Transition
      1. Demographic Transition - Traditionally, economic development, through stimulating the demographic transition, was thought the best population control strategy. Industrialization will not be the answer, though, where energy, economic and natural resources, and time are limited. There are four reasons economic development might not work:
         a. Economic resources of developing nations may be too limited.
         b. Demographic transition takes a lot of time.
         c. Population growth can outstrip economic growth.
         d. Fossil fuel energy sources are diminishing and becoming more expensive and could be economically unavailable to developing nations.
   B. Family Planning and Population Stabilization
      1. These programs may be voluntary, extended voluntary, or forced and aim to provide birth control and motivation for its use to all couples.
   C. Small-Scale Sustainable Economic Development, Better Health Care, and Improvements in the Status of Women
      1. Such initiatives aim to create incentives, such as jobs and improved health care, for women to have fewer children.

D. Sustainable Populations in the Developed Countries
   1. Though growth in developed countries is comparatively slow, their high level of per capita consumption is important. A single American or Canadian uses 20 to 40 times more resources than a citizen of the developing world.
   2. Financial assistance and sharing of appropriate technology with less developed countries may be the most effective way for developed nations to assist other countries in need of aid.
E. Creating Sustainable Populations in the Developing Countries
   1. Improved funding and implementation of population control programs in developing countries is urgently needed.

III. Overcoming Barriers
   A. Psychological and Cultural Barriers - Population control efforts must cross the psychological and cultural barriers of security and esteem tied to large family size.
   B. Educational Barriers - Less educated people tend to have more children than do the more highly educated.
   C. Religious Barriers - Religious doctrines continue to foster excess reproduction among those who uncritically accept religious dogma.

IV. Ethics and Population Stabilization
   A. Is Reproduction a Personal Right? Many argue that the collective rights of all living humans, and the integrity of the biosphere which those rights presuppose, supersede individuals' rights to reproduce excessively.
   B. Is It Ethical Not to Control Population Growth? Ethical obligations to future generations suggest that it is not morally right to allow population growth today at the expense of tomorrow's citizens and other species.

V. Status Report
   A. Major efforts are underway to reduce the growth of the human population. World population growth has dropped substantially in the last 25 years.
   B. Growth rates in the developed nations have fallen in the last two decades.
   C. Most of the new growth in the coming decades will occur in the less developed nations.

Suggestions for Presenting the Chapter

- Instructors might look for local examples of urban sprawl for class observation and study. How is your city dealing with urban development? What are the sustainable solutions to your local problems?
- A speaker on family planning will provide interesting topics for class discussion.
- Instructors might assign individual or groups the task of studying a particular countries' population issues. Compare the problems encountered in developed and developing countries.
- Instructors should explore the ethical issues surrounding population control. This can be done by class/group discussion, writing assignments and speakers.
- Recommended web sites:

Zero Population Growth: http://www.zpg.org
Negative Population Growth: http://www.npg.org/

U.S. Census Bureau: http://www.census.gov/main/www/popclock.html
United Nations Fund for Population Activities: http://www.unfpa.org/
U.S. Agency for International Development: http://www.usaid.gov/

True/False Questions

1. ___ Economic development can be a powerful force for reducing population growth.
2. ___ Family planning measures permit couples to have more children and are vital to reaching a sustainable population level.
3. ___ Birth control includes devices or methods that reduce births in a population.
4. ___ Abstinence is a technique used to increase human fertility.
5. ___ Contraceptives are methods that prevent sperm and egg from uniting.
6. ___ Family planning allows couples to determine the number and spacing of offspring.
7. ___ The United States has a forced family planning program.
8. ___ Raising the standard of living of the world's poor is an essential component of a multifaceted population program.
9. ___ The U.S. Agency for International Development has opposed the development of family planning programs in the developing world.
10. ___ The United Nations Fund for Population Activities is a leader in promoting family planning throughout the world.
11. ___ In many developing countries, children are seen as an asset to their parents and childbearing enhances a woman's status.
12. ___ The Catholic religion forbids all "unnatural" methods of birth control such as the pill, the condom, the diaphragm and abortion.
13. ___ To many people, reproduction is a basic human right.
14. ___ Preindustrial societies exhibit low birth rates and low death rates.
15. ___ Postindustrial societies exhibit high birth rates and high death rates.
16. ___ The intentional interruption of pregnancy through surgical means or drug treatments is called induced abortion.
17. ___ Family planning programs promoted by governments are called voluntary programs.
18. ___ The state of Oregon is a world leader in urban growth management.
19. ___ Three primary barriers lie in the way of achieving a sustainable human population: psychological and cultural, educational and religious.
20. ___ Poverty is one of the root causes of rampant population growth.

True/False Key:

1. T 2. F 3. T 4. F 5. T 6. T 7. F 8. T 9. F 10. T 11. T 12. T 13. T 14. F 15. F 16. T 17. F 18. T 19. T 20. T

Fill-in-the-Blank Questions

1. _____ is refraining from intercourse.
2. Family planning programs promoted by governments are called extended _____ programs.
3. _____ planning allows couples to determine the number and spacing of offspring.
4. _____ are any chemicals, devices or methods that prevent sperm and egg from uniting.
5. _____ family planning programs involve strict governmental limitations on family size.
6. The _____ stage of demographic transition is characterized by high birth rates and high death rates.
7. The use of _____ measures are devices or techniques that reduce the chance of fertilization.
8. One of the root causes of rampant population growth is the _____ of women.
9. _____ programs make birth control information and methods available to the public at a low cost.
10. The transitional stage of demographic transition is characterized by _____ birth rates and falling death rates.
11. The intentional interruption of pregnancy through surgical means or drug treatments is called induced _____.
12. _____, the largest nation on earth, has a forced family planning program.
13. A world leader in urban growth management is the state of _____.
14. The U.S. Agency for International _____ was the major sponsor of family planning programs in the developing world from 1965 to 1980.
15. Ecologist Garrett Hardin argues that the integrity of the biosphere and the Earth's _____ capacity should be the guiding principle in the debate.
16. Today, one-third of the world's population is under the age of __ and is soon to enter its reproductive years.
17. It took Finland over _____ years to approach a balance between birth rate and death rate.
18. Planned Parenthood in the United States is a private, _____ organization with clinics in large cities.
19. The demographic changes that occur with economic development in developing nations is called demographic _____.
20. Studies show that a 1% growth in the labor force requires a __% economic growth rate.

Fill-in the Blank Key:

1. abstinence
2. voluntary
3. family
4. contraceptives
5. forced

6. preindustrial
7. contraceptive
8. inequality
9. voluntary
10. high
11. abortion
12. China
13. Oregon
14. development
15. carrying
16. 15
17. 200
18. nonprofit
19. transition
20. 3

Multiple Choice Questions

1. About _____ of the world's people live in extreme poverty.
    a. 100,000
    b. 1,000,000
    c. 10,000,000
    * d. 1,000,000,000
    e. 10, 000,000,000

2. Which of the following responses is **not** one of the root causes of rampant population growth?
    a. poverty
    b. lack of education
    c. poor health care
    * d. family planning
    e. inequality of women

3. Contraceptives are birth control measures that use _____ to prevent pregnancy.
    a. chemicals
    b. methods
    c. devices
    d. a and b only
    * e. a, b, and c

4. Family planning programs not promoted by the government are:
    * a. voluntary programs
    b. extended voluntary programs
    c. forced family planning programs
    d. a and b only
    e. b and c only

5. Extended voluntary family planning programs:
    a. are mandatory.
    * b. are government sponsored.
    c. are privately sponsored.
    d. do not provide financial aid.
    e. set a limit on the number of children.

6. Collective rights is a concept that opposes:
    a. abortion
    b. family planning
    c. contraception
    * d. individual rights
    e. population stabilization

7. Forced family planning is found in which of the following nations?
   a. Thailand
   b. Sri Lanka
   * c. China
   d. Australia
   e. Brazil

8. Which state is a world leader in growth management?
   a, Wisconsin
   b. Colorado
   c. California
   * d. Oregon
   e. Maine

9. A single American or Canadian uses _____ times more resources than a citizen of the developing world.
   a. 5 to 10
   b. 10 to 20
   * c. 20 to 40
   d. 50 to 60
   e. 70 to 80

10. Environmental Impact = _____ times per capita consumption times pollution and resource use per unit of consumption.
   a. reproductive rate
   b. total fertility rate
   * c. population size
   d. growth rate
   e. crude birth rate

11. The concept that individual rights should not take precedence when the welfare of society is threatened is called:
   a. collective bargaining
   b. double jeopardy
   * c. collective rights
   d. tort
   e. natural law

12. Which of the following are population control strategies for developed countries?
   a. Stabilize population growth by fostering sex education.
   b. Provide financial assistance to developing countries.
   c. Provide assistance to population control programs.
   d. Make trade with less developed countries equitable and freer.
   * e. all of the above

13. How much money is donated to developing nations for family planning from outside sources?
   a. $200 million
   b. $400 million
   * c. $600 million
   d. $800 million
   e. $900 million

14. In developing countries, children are often seen as an asset to their parents, and childbearing enhances a woman's social status. These barriers to population control are primarily:
   * a. psychological and cultural
   b. educational
   c. religious
   d. social
   e. political

15. The higher the level of education in a population, the lower its:
   a. family planning services.
   b. environmental damage.
   * c. fertility rate.
   d. marriage rate.
   e. resource use.

16. In countries where family planning is culturally unacceptable, family planning can be promoted as a way to:
   a. reduce family costs.
   * b. space births to improve maternal/child health.
   c. increase working families.
   d. reduce welfare payments.
   e. increase infant mortality.

17. During the last 25 years the world's population growth rate has:
    a. increased.
    * b. decreased.
    c. not changed.
    d. not been determined.
    e. too many variables to be estimated.

18. During the past 25 years the population growth rate in developing nations has:
    a. increased.
    * b. decreased.
    c. not changed.
    d. not been determined.
    e. too many variables to be estimated.

19. Global population growth is now at ____%.
    a. 0.5
    b. 1.0
    * c. 1.5
    d. 2.0
    e. 2.5

20. Which of the following countries/regions has **not** reduced its population growth rate?
    * a. Africa
    b. China
    c. Korea
    d. Taiwan
    e. Costa Rica

21. Which stage of demographic transition is characterized by stable population size and high birth and death rates?
    * a. Stage 1
    b. Stage 2
    c. Stage 3
    d. Stage 4
    e. Stage 5

22. Abstinence refers to birth control by:
    a. using condoms.
    b. using the rhythm method.
    * c. refraining from intercourse.
    d. oral contraceptives.
    e. induced abortion.

23. Which of the following statements is not valid?
    a. Population stabilization is essential to achieve a sustainable world.
    b. Family planning can promote reduced birth rates and lower population growth.
    * c. All consumption patterns can be made sustainable through conservation and other means.
    d. One third of the world's population is under the age of 15 and is soon to enter its reproductive years.
    e. Major efforts are underway to slow the growth of the human population.

24. Experts in population suggest that measures should be taken to enhance women's:
    a. rights
    b. responsibilities
    c. social standing
    d. educational opportunities
    * e. all of the above

25. Sustainable growth is a concept that:
    a. depends on technological improvement.
    b. believes economic growth can continue by applying energy efficiency.
    c. suggests we can continue to grow both our economy and population in a sustainable way.
    d. is certainly unattainable.
    * e. all of the above.

26. Which of the following statements outlines the two basic challenges to achieving a sustainable human population?
    a. Use sustainable growth and technology.
    * b. Slow population growth and eventually reduce population size.
    c. Support family planning and increased efficiency.
    d. Slow population growth and encourage conservation.
    e. Encourage mandatory family planning and promote energy efficiency

27. Which stage of demographic transition is characterized by improvements in health care and sanitation , falling death rates and population growth?
    a. Stage 1
    * b. Stage 2
    c. Stage 3
    d. Stage 4
    e. Stage 5

28. Which stage of demographic transition is characterized by stable population size and low birth and death rates?
    a. Stage 1
    b. Stage 2
    c. Stage 3
    * d. Stage 4
    e. Stage 5

29. Which stage of demographic transition is characterized by declining birth rates and economic development?
    a. Stage 1
    b. Stage 2
    * c. Stage 3
    d. Stage 4
    e. Stage 5

30. Each birth averted by family planning yields a savings of between _____ per year in social services.
    a. $5 and $15
    b. $10 and $20
    * c. $15 and $200
    d. $100 and $500
    e. $1000 and $5000

31. Which of the following organizations recently made substantial investments in clothing factories that will employ Egyptian women?
    a. U.S. Agency for International Development
    b. International Planned Parenthood Federation
    * c. United Nations Fund for Population Activities
    d. Amnesty International
    e. The World Bank

32. The use of chemicals, devices or methods that prevent sperm and egg from uniting are called:
    a. abstinence
    * b. contraceptives
    c. induced abortion
    d. sterilization
    e. extended voluntary family planning

33. Government agencies handing out information on birth control and sterilization constitute _____ family planning.
    a. voluntary
    * b. extended voluntary
    c. forced
    d. government
    e. sustainable

34. What is the rationale for concentrating research on the social, cultural, and psychological aspects of reproduction?
    a. A higher standard of living and increased job opportunities could result.
    * b. What is needed is more motivation for population control.
    c. The rich-poor gap could narrow decreasing social strife.
    d. This will stimulate economic growth and reduce poverty.
    e. Freer trade will increase per capita income and raise standards of living.

35. Which of the following responses is a reason to question demographic transition as a way to slow population growth in developing countries?
    a. Economic resources of many developing countries cannot support the industrial development needed for demographic transition.
    b. Demographic transition may take too long too be effective in developing countries considering their current doubling times.
    c. Population growth can outstrips economic growth.
    d. Fossil fuel resources that were essential to the demographic transition of developed countries may not be available since supplies are diminishing.
    * e. all of the above

36. Which of the following responses is **not** a strategy to control population in developing countries?
*   a. Seek programs that concentrate wealth in the hands of a few wealthy land owners.
    b. Seek funding from the United Nations and developed countries.
    c. Seek to change cultural taboos against birth control.
    d. Finance education in rural regions emphasizing birth control.
    e. Develop an effective national plan to encourage family planning and the availability of contraception.

37. Recent studies show that it costs low income families in the U.S. about _____ to raise a child.
    a. $20,000
    b. $30,000
    c. $40,000
    d. $50,000
*   e. $60,000

38. Middle income families in the U.S. Spend about _____ to raise each child.
    a. $70,000
    b. $80,000
*   c. $90,000
    d. $100,000
    e. $120,000

39. The childbearing years are:
    a. 10 to 27.
    b. 12 to 32.
*   c. 15 to 44.
    d. 18 to 58
    e. 20 to 60

40. People without educational opportunities:
    a. tend to marry younger.
    b. pursue careers that do not interfere with childbearing.
    c. have a harder time learning about the proper use of contraceptives.
    d. find it more difficult to learn about alternatives to childbearing.
*   e. all of the above

11

Creating a Sustainable System of Agriculture to Feed the World's People

Chapter Outline

Hunger, Malnutrition, Food Supplies and the Environment
Hunger, Poverty, and Environmental Decay
Declining Food Supplies
The Challenge Facing World Agriculture: Feeding People/Protecting the Planet

Understanding Soils
What is Soil?
How Is Soil Formed?
What Is a Soil Profile?

Barriers to a Sustainable Agricultural System
Soil Erosion
Desertification
Farmland Conversion
Declines in Irrigated Cropland
Waterlogging and Salinization
Declining Genetic Diversity in Crops and Livestock

Solutions: Building a Sustainable Agricultural System
Protecting Existing Soil and Water Resources
Soil Enrichment Programs
Increasing the Amount of Land in Production
Increasing the Productivity of Existing Land: Developing Higher-Yield Plants and Animals
Protecting Wild Plant Species: Habitat Protection and Germ Plasm Repositories
Developing Alternative Foods: Native Species as Sustainable Food Sources
Fish from the Sea and Aquaculture
Eating Lower on the Food Chain
Reducing Pest Damage and Spoilage
Legislation and New Policies: Political and Economic Solutions

Key Terms

sustainable agriculture	undernourished	malnourished
infectious diseases	kwashiorkor	marasmus
soil	parent material	soil profile
horizons	litter layer	topsoil
humus	subsoil	soil erosion
natural erosion	accelerated erosion	desertification
farmland conversion	waterlogging	water table
salinization	Green Revolution	subsidies
minimum tillage	herbicides	contour farming
strip cropping	terracing	shelterbelts
organic fertilizers	green manure	synthetic fertilizer
crop rotation	hybrids	selective breeding
genetic engineering	germ plasm repositories	overfishing
fish farms	aquaculture	mariculture

Objectives

1. Define the following terms: "undernourished", "malnourished", "kwashiorkor", and "marasmus".
2. Discuss how hunger and malnutrition are related to environmental degradation.
3. Discuss the trend in per capita grain production and its ramifications for human populations.
4. Define the following terms: "soil", "parent material", "topsoil", "litter layer", "humus", and "subsoil".
5. List the five major soil horizons and their common names.
6. Define the following barriers to sustainable agriculture and describe their impact on the agricultural system: "soil erosion", "desertification", "farmland conversion", "declines in irrigated cropland", "waterlogging", "salinization".
7. Discuss the successes and failures of the Green Revolution.
8. List the techniques used in soil conservation.
9. Discuss how water can be used more efficiently on irrigated croplands.
10. Discuss the different types of soil enrichment used today and their advantages.
11. Discuss the techniques used to increase productivity of crops and livestock.
12. Discuss the importance of maintaining genetic diversity in food crops and in our genetic repositories.
13. Summarize the sources of alternative foods available to human populations.
14. Define the following terms: "overfishing", "fish farms", "aquaculture", and "mariculture".
15. Discuss why eating lower in the food chain could provide more food for the world's human population.
16. Discuss the significance of food loss by pest damage and spoilage.

Lecture Outline

I. Hunger, Malnutrition, Food Supplies, and the Environment - Hunger is widespread; it is especially prevalent in Asia, Africa, and Latin America. Large percentages of the populations of poor countries suffer from some combination of malnourishment and undernourishment.

   A. Diseases of Malnutrition - Millions of people die yearly from malnutrition and undernourishment.

      1. Kwashiorkor - Protein-deficiency disease, or kwashiorkor, strikes children soon after weaning.

      2. Marasmus - Protein- and calorie-deficiency produce marasmus. Less obvious effects of moderate malnutrition include decreased immunity.

   B. Hunger, Poverty, and Environmental Decay - Severe malnutrition produces serious and permanent mental and physical impairment.

   C. Declining Food Supplies - Global warming and soil erosion/deterioration have led to declining per capita food production over the past decade.

   D. The Challenges Facing World Agriculture: Feeding People/Protecting the Planet - A sustainable system of agriculture is needed to meet the present and future needs for food and to protect soil and water.

II. Understanding Soils

   A. What is Soil? Soils are mixtures of organic and inorganic materials varying in a number of features.

   B. How Is Soil Formed? A variety of physical processes and organisms contribute to soil formation from parent material.

   C. What is a Soil Profile? Soils have layers which differ in type and thickness between different soil types. Climate, geological feature,. biotic factors, and age determine a given region's soil profile.

III. Barriers to a Sustainable Agricultural System

   A. Soil Erosion - This is the most serious agricultural problem today.

      1. Accelerated erosion largely results from human activities such as overgrazing or unsustainable cultivation practices. Accelerated erosion not only destroys productivity but contributes heavily to air and water pollution problems.

      2. Natural erosion occurs in areas in the absence of human intervention.

   B. Desertification: Turning Cropland to Desert - Global climate change and mistreatment of soils in arid regions leads to desertification. Previous civilizations have caused desertification; today, the problem is widespread, but particularly bad in Africa.

   C. Farmland Conversion - Farmland is lost as urbanization, energy production, transportation, and other forms of development take land out of agricultural production; this is known as farmland conversion.

   D. Declines in Irrigated Cropland - The amount of irrigated cropland per capita is on the decline. Measures that increase the efficiency of water use may prove helpful in providing an adequate supply of irrigation water.

   E. Waterlogging and Salinization

      1. Waterlogging occurs when too much water is applied to a crop. Worldwide about one-tenth of the irrigated cropland suffers from waterlogging.

      2. Salinization is the accumulation of salts and minerals in the soil that occurs after irrigation. Evaporation will concentrate the salts and minerals in the soil over

time and reduce productivity. About one-fourth of the irrigated farmland worldwide suffers from salinization.

  F. Declining Genetic Diversity in Crops and Livestock
1. Green Revolution - New varieties of crops, such as those introduced in the Green Revolution, have substantially increased yield but decreased genetic diversity of crop plants; this increases vulnerability to pests, disease organisms, and environmental stress.
2. Habitat Destruction - Local varieties that are well adapted to regional sites are at risk of loss because of lack of use and propagation. This is the largest source of genetic crop diversity. Monoculture has increased the risk of large losses in productivity as a result of pest infestation.

  G. Politics, Agriculture, and Sustainability
1. Government policies have not always fostered sustainable use of agricultural systems. Subsidies have contributed to unsustainable agricultural practices and have just been recently changed. Laws and policies must be systematically examined and revised with global sustainability in mind.

IV. Solutions: Building a Sustainable Agricultural System - Creating a sustainable agricultural system will require a multifaceted approach, including measure to slow and perhaps stop human population growth.

  A. Protecting Existing Soil and Water Resources
1. Soil conservation: six strategies for the basis for sustainable practices to control soil erosion.
    a. Minimum Tillage or Conservation Tillage
    b. Contour Farming
    c. Strip Cropping
    d. Terracing
    e. Gully Reclamation
    f. Shelterbelts
2. Overcoming the Economic Obstacles to Soil Erosion Controls
    a. Government incentives for farmers to practice good sustainable agriculture.
    b. The 1985 farm bill was a step in the right direction but concerns are that it may be diluted by further legislation.
3. Preventing Desertification
4. Reducing Farmland Conversion
5. Saving Irrigated Cropland/Using Water More Efficiently
6. Preventing Salinization and Waterlogging

  B. Soil Enrichment Programs - Farming reduces valuable soil nutrients. These can be replaced by sustainable agricultural techniques.
1. Organic Fertilizers
2. Synthetic fertilizers only partially replenish the soil.
3. Crop Rotation

  C. Increasing the Amount of Land in Production
1. Development of new varieties of plants and animals.
    a. Selective breeding
    b. Genetic engineering

  D. Protecting Wild Plant Species - Habitat Protection and Germ Plasm Repositories
1. Habitat protection
2. Seed banks

  E. Developing Alternative Foods: Native Species as Sustainable Food Sources - Many native plant and animal species would be used to provide food. Native animals offer many benefits over domestic livestock, including their resistance to disease-causing organisms.

F. Fish from the Sea and Aquaculture
    1. Many of the world's commercially important fish stocks are in danger of being depleted.
    2. Global efforts are needed to preserve current fish stocks.
    3. Commercial fish farming could provide additional food for a growing population.
G. Eating Lower on the Food Chain
    1. Developing nations should concentrate on grain production rather than meat.
    2. A change in eating habits in the western world could increase the amount of food available for human consumption.
H. Reducing Pest Damage and Spoilage
    1. About 30% of all agricultural output is destroyed by pests, spoilage, and diseases.
    2. Improvements in storage and transportation can reduce food loss.
I. Legislation and New Policies: Political and Economic Solutions
    1. Solving world hunger and creating a sustainable system of agriculture will require dramatic changes in government policies worldwide.

Suggestions for Presenting the Chapter

- Instructors should emphasize that adequate nutrition for the world's peoples is not just a social or moral issue. Nutrition has important biological ramifications in human ecology and will directly or indirectly affect the environment.
- Instructors should emphasize that the dependency on technology fostered by the Green Revolution is not the ultimate solution to our agricultural problems. The movement to sustainable production must be made if the people of the world are going to be fed and the land is going to remain productive for future generations. The goal should be long-term sustainable production not short-term profit.
- Instructors are encouraged to investigate local agricultural practices and inform the students about them. This can be achieved by field trips, speakers, and class discussion.
- Visiting local supermarkets and examining the sources of produce and the availability of organically produced fruits, vegetables, and meat is an informative class project. What foods available in your local stores are produced locally? What foods are organically produced? Why should we seek to purchase organically produced foods? What are important local sources of organic food?
- Explore the topic of organic gardening with your students. How many students participate in gardening? Does your institution have a garden plot? What are the advantages of growing your own food? Why is organic gardening a sustainable activity?
- Recommended web sites:

  Food and Agriculture Organization of the United Nations: http://www.fao.org/
  UDSA Agencies and Programs: http://www.usda.gov/services.html
  Worldwatch Institute: http://www.worldwatch.org/
  Desertification - Monitoring and Forecasting: http://www.planetary.caltech.edu/~arid/
  Marine Biotechnology & Aquaculture Opportunities:
  http://www.nalusda.gov/bic/bio21/aqua.html

Multiple Choice Questions

1. People who do not get sufficient calories in their diet are:
   a. malnourished
   * b. undernourished
   c. incapacitated
   d. starved
   e. stressed

2. Malnourishment refers to lacking adequate _____ in the diet.
   a. water
   * b. nutrients and vitamins
   c. fiber
   d. bulk
   e. calories

3. Severe protein deficiency can cause _____:
   a. scurvy
   b. malaria
   * c. kwashiorkor
   d. hypertension
   e. bulimia

4. Calorie and protein deficiency can cause:
   * a. marasmus
   b. rickets
   c. malaria
   d. hypertension
   e. scurvy

5. Long lasting effects of malnutrition may be:
   * a. mental retardation
   b. atherosclerosis
   c. lung cancer
   d. obesity
   e. hypertension

6. Which of the following responses is one of the four components of soil?
   a. water
   b. air
   c. inorganic materials
   d. organic matter
   * e. all of the above

7. The world population is growing at ____ per day.
   a. 15,000
   b. 26,000
   c. 73,000
   d. 160,000
   * e. 240,000

8. Between 1950 and 1970 improvements in agriculture increased world per capita grain production ____%.
   a. 10
   b. 20
   * c. 30
   d. 40
   e. 50

9. Between 1984 and 1989, food production per capita fell ___%.
   a. 4
   b. 8
   c. 12
   * d. 14
   e. 20

10. Which of the following responses is a factor that contributed to the decline in per capita food production between 1984 and 1989?
    a. warming global climate
    b. soil erosion
    c. soil deterioration
    d. population growth
    * e. all of the above

11. The layers of different color and composition of a soil are called the:
    a. soil strata
    b. soil sample
    * c. soil profile
    d. soil test
    e. soil compaction

12. The inorganic materials in soil do **not** include:
  * a. leaves
  b. clay
  c. silt
  d. sand
  e. gravel

13. The organic matter in soil do **not** include:
  a. leaves
  b. roots
  * c. clay
  d. bacteria
  e. earthworms

14. The type of rock that is the basis for soil formation is called the:
  a. strata
  b. bedrock
  * c. parent material
  d. substrate
  e. block

15. It may take 200 to 1200 years to form _____ of soil from hard rock, depending on the climate.
  a. 0.5 inches
  * b. 1.0 inches
  c. 2.0 inches
  d. 3.0 inches
  e. 4.0 inches

16. Which of the following is one of the factors determining the type of soil that forms?
  a. climate
  b. parent material
  c. biological organisms
  d. topography
  * e. all of the above

17. _____ erosion occurs in areas without the activities of our species:
  a. accelerated
  * b. natural
  c. rill
  d. gully
  e. sheet

18. The uppermost layer of the soil is the:
  a. A horizon
  b. B horizon
  c. C horizon
  d. D horizon
  * e. O horizon

19. This soil layer is a transition zone between the parent material and soil layers above; it is the:
  a. A horizon
  b. B horizon
  * c. C horizon
  d. D horizon
  e. O horizon

20. This layer is the rock from which soils are formed:
  a. A horizon
  b. B horizon
  c. C horizon
  * d. D horizon
  e. O horizon

21. This layer of the soil is the rich in inorganic and organic materials and supports crop production:
  * a. A horizon
  b. B horizon
  c. C horizon
  d. D horizon
  e. O horizon

22. This soil layer is also known as the zone of accumulation because it collects minerals and nutrients leached from above.
  a. A horizon
  * b. B horizon
  c. C horizon
  d. D horizon
  e. O horizon

23. Each year in the United States, _____ acres of rural farmland are lost every day to other uses.
  a. 1500
  b. 2500
  * c. 3500
  d. 4500
  e. 6000

24. The World Resources Institute estimates off-site damage from soil erosion in the United States at over _____ a year.
   a. $2 billion
   b. $4 billion
   c. $6 billion
   d. $8 billion
   * e. $10 billion

25. The process where cropland, rangeland, and pasture become too dry to use because of climate change and poor land management is called:
   a. climatic transition
   b. succession
   * c. desertification
   d. leaching
   e. salinization

26. Which of the following is a potential cause of desertification?
   a. drought
   b. global warming
   c. overgrazing
   d. deforestation
   * e. all of the above

27. The use of valuable farmland for nonagricultural activities is called:
   * a. farmland conversion
   b. desertification
   c. deforestation
   d. strip development
   e. crop rotation

28. The farming technique that reduces the amount of land disturbed during soil preparation before planting is called:
   * a. conservation tillage
   b. conventional tillage
   c. moldboard plowing
   d. crop rotation
   e. windbreaks

29. Globally about ___% of the world's cropland is irrigated.
   a. 4
   b. 8
   * c. 18
   d. 36
   e. 72

30. Irrigation must be done carefully because it can cause:
   a. wind erosion and fertilization
   b. drought and desertification
   c. insect infestation and fungal growth
   * d. waterlogging and salinization
   e. lowered water tables and toxic pollution

31. The Green Revolution was a worldwide effort to improve the:
   a. soils of the world.
   b. industrial pollution policies.
   * c. agricultural productivity.
   d. organic agriculture of developing nations.
   e. soil erosion problem.

32. Using only a few hybrid strains of grain and the technique of monoculture, modern agriculture is faced with this important problem:
   a. The high costs of new hybrid crops.
   b. Development of new hybrids is technically difficult.
   * c. We are contributing to the loss of genetic diversity.
   d. Monoculture requires too much irrigation.
   e. Hybrid grains have low productivity.

33. New varieties of plants produced by selective breeding of closely related plants are called:
   a. mutants
   b. genetically engineered plants
   * c. hybrids
   d. ecotypes
   e. inbred varieties

34. Which of the following responses are advantages of minimum tillage?
    a. reduced soil erosion
    b. reduced energy consumption
    c. reduced use of pesticides
    d. reduced soil compaction
 * e. all of the above

35. Planting crops perpendicular to the slope of the land is called:
    a. strip cropping
 * b. contour farming
    c. terracing
    d. gully reclamation
    e. crop rotation

36. Earthen embankments placed across the slope to check erosion are characteristic of :
    a. strip cropping
    b. contour farming
 * c. terracing
    d. gully reclamation
    e. crop rotation

37. Trees planted along the perimeter of crops to reduce wind erosion are called:
    a. wind rows
 * b. shelterbelts
    c. waterways
    d. reforestation
    e. riparian vegetation

38. The use of cover crops that are plowed under to increase the fertility of the soil is called:
 * a. green manure
    b. synthetic fertilizer
    c. soil enhancement
    d. crop rotation
    e. no-till agriculture

39. The process of transferring genes into the genetic material of an organism is:
    a. mutation
 * b. genetic engineering
    c. natural selection
    d. selective breeding
    e. cloning

40. Fish farms are forms of aquatic agriculture called:
    a. hydroponics
 * b. aquaculture
    c. irrigation
    d. commercial fishing
    e. subsistence fishing

12
Preserving Biological Diversity

Chapter Outline

Biodiversity: Signs of Decline

Causes of Extinction and the Decline in Biodiversity
    Physical Alteration of Habitat
    Commercial Hunting and Harvesting
    The Introduction of Foreign Species
    Pest and Predator Control
    The Collection of Animals and Plants for Human Enjoyment, Research, and Other Purposes
    Pollution
    Biological Factors That Contribute to Extinction
    The Loss of Keystone Species
    A Multiplicity of Factors

Why Protect Biodiversity?
    Aesthetics and Economics
    Food, Pharmaceuticals, Scientific Information, and Products
    Protecting Free Services and Saving Money
    Ethics - Doing the Right Thing

How to Save Endangered Species and Protect Biodiversity - A Sustainable Approach
    Protecting All Species
    Stopgap Measures: First Aid for an Ailing Planet
    Long-term Preventive Measures
    Personal Solutions

Key Terms

natural extinction	accelerated extinction	endangered species
threatened species	sport hunting	commercial hunting
subsistence hunting	poaching	keystone species
ecotourism	biodiversity	Endangered Species Act
debt-for-nature swap	buffer zones	wildlife corridor
extractive reserve	predator control	

Objectives

1. Define the following terms: "natural extinction", "accelerated extinction", "endangered species", and "threatened species".
2. Discuss the causes of extinction and the decline of biodiversity.
3. List the reasons to protect biodiversity.
4. Discuss the significance of the Endangered Species Act.
5. Summarize the measures used to save endangered species and protect biodiversity.
6. Describe the "debt-for-nature swap" option and discuss where this technique has been used to save biodiversity.
7. Define the following terms: "ecological islands", "buffer zones", "wildlife corridor", and "extractive reserve".
8. Suggest some actions you can take personally to foster biodiversity.

Lecture Outline

I. Biodiversity: Signs of Decline - Extinction is a natural process, but our activities have accelerated the rate of species extinction to levels which are neither natural nor desirable.

II. Causes of Extinction and the Decline in Biodiversity
   A. Physical Alteration of Habitat
      1. Habitat alteration/destruction is the primary cause of species extinction today.
      2. Tropical rainforests comprise the most rapidly disappearing habitat; other threatened, critical habitats include coral reefs, wetlands, estuaries, tall grass prairie, and temperate rainforest.
   B. Commercial Hunting and Harvesting - Hunting for profit threatens certain species which are commercially valuable, such as whales, elephants, and large cats.
   C. The Introduction of Foreign Species - Foreign species can harm native species through the direct effects of competition and through indirect effects.
   D. Pest and Predator Control - Pesticides can harm many non-target species. Some species of predators have been seriously damaged by misguided eradication efforts aimed at protecting livestock.
   E. The Collection of Animals and Plants for Human Enjoyment, Research, and Other Purposes- Collecting animals to serve as research specimens or as pets depletes wild populations; trade in wild plants and plant parts has decimated many populations of certain species, especially the cacti and insectivorous plants.
   F. Pollution - A wide variety of pollutants threaten wildlife directly and make them more susceptible to disease and environmental stress. Sources include agricultural runoff/drainage, acid rain, global warming, ozone depletion, and various industrial toxins.
   G. Biological Factors That Contribute to Extinction - Certain traits make some species especially vulnerable to extinction. These include a high critical population size, specialization, narrow range, large size, and intolerance of human presence.
   H. The Loss of Keystone Species - Keystone species, though often inconspicuous or obscure, play a critical role in maintaining ecosystem balance and integrity. Their protection is particularly important.

I. A Multiplicity of Factors - Many of the factors discussed previously may interact to make species loss more probable.

III. Why Protect Biodiversity? Arguments for protecting endangered species and preserving biodiversity can be made on utilitarian and ethical grounds.
   A. Aesthetics - Each species has some aesthetic value; every extinction diminishes the aesthetic quality of human life.
   B. Food, Pharmaceuticals, Scientific Information, and Products - A large number of species do or potentially might contribute to human utility; that is, they directly or indirectly benefit us through the provision of health products or scientific information.
   C. Protecting Free Services and Saving Money - Natural systems provide billions of dollars of benefits to human societies.
   D. Ethics -If, as some believe, every species has a right to exist, then humans have an ethical obligation to respect that right.

IV. How to Save Endangered Species and Protect Biodiversity
   A. Protecting All Species - Many ecologically important species could face extinction if efforts are not broadened. Currently, the focus is on saving the most appealing or most visible species.
   B. Stopgap Measures: First Aid for an Ailing Planet
      1. The Endangered Species Act - The Endangered Species Act is the major legal instrument for species preservation in the U.S..
      2. Worldwide, weak laws and inadequate enforcement make species protection more difficult.
      3. Captive Breeding Programs - Zoos Lend a Hand
         a. Zoo-sponsored captive breeding programs may help save a few critically endangered species.
   C. Long-Term Preventive Measures - The following are long-term steps to protect biodiversity:
      1. Restructuring Human Systems for Sustainability - Protecting biodiversity will be aided by addressing the root causes of the crisis of unsustainability: our inefficient use of resources, runaway population growth, etc.
      2. Setting Aside Biologically Rich Regions
      3. Buffer Zones and Wildlife Corridors: Protecting and Connecting Vital Areas
      4. Extractive Reserves
      5. Improving Wildlife Management
   D. Personal Solutions - Saving species and protecting biodiversity require personal action. The following are some suggestions:
      1. Use only what you need, and use all resources efficiently.
      2. Recycle and buy recycled products.
      3. Use renewable resources; support government programs aimed at increasing their use.
      4. Help restore ecosystems; support groups that take an active role in these efforts.
      5. Limit your family size; support private and government efforts throughout the world to provide family planning services and other means to help slow the growth of the human population.

Suggestions for Presenting the Chapter

- Instructors should review the loss of biodiversity in your state or local area.
- Instructors might involve their classes in restoration projects which focus on habitat improvement to foster biodiversity.
- A review of state sponsored programs to encourage biodiversity often proves to be topical to students. A speaker from your state department of natural resources or a trip to public land in your area often provide good opportunities to explore issues of biodiversity and habitat destruction/preservation.
- Many videos are available on the plight of the tropical rain forest. These can be viewed during regular class time or given as an assignment for viewing outside of class time
- Recommended web sites:

    Endangered Species Program/U.S. Fish and Wildlife Service: http://endangered.fws.gov/
    The Wildlands Project: http://www.wildlandsproject.org/
    Biodiversity & Worldmap:
    http://www.nhm.ac.uk/science/projects/worldmap/worldmap/demo2.htm
    Address & slide show by E.O. Wilson:
    http://www.saveamericasforests.org/news/EOWilsonIntro.htm
    Yellowstone to Yukon Conservation Initiative: http://www.rockies.ca/y2y/

True/False Questions

1. ___ Accelerated extinction is the loss of species as a result of human activities.
2. ___ An endangered species is one that is still abundant in its natural range but is likely to become extinct in time.
3. ___ Habitat alteration is the number one cause of species extinction.
4. ___ Subsistence hunting includes large-scale, commercial, whale hunting and hunting African rhinos for their horns.
5. ___ Illegal hunting is called poaching.
6. ___ In Hawaii, 90% of all bird species have been wiped out by human inhabitants and organisms that humans have introduced.
7. ___ Keystone species are organisms upon which many other species in an ecosystem depend.
8. ___ The Passenger Pigeon is now a common species thanks to the Endangered Species Act.
9. ___ Protecting natural systems helps preserve many ecological services such as flood control and water pollution abatement.
10. ___ The United States Congress passed the Endangered Species Act in 1914.
11. ___ A wildlife corridor is a strip of land that connects habitats set aside to protect species.
12. ___ A buffer zone is a region of intense human development, for instance, timber cutting and manufacturing, that is located near a wilderness area.
13. ___ An extractive reserve is land set aside for native people to use on a sustainable basis.

14.___ The first extractive reserve established in the world is in the Arctic National Wildlife Refuge in Alaska.
15.___ The Endangered Species Act requires the U.S. Fish and Wildlife Service to develop recovery plans for endangered species.
16.___ Tropical rain forests house at least half of the Earth's species.
17.___ DDT was the chemical responsible for the restoration of peregrine falcon populations in the United States.
18.___ 50% of all prescription and nonprescription drugs are made with chemicals derived from or originally extracted from wild plants.
19.___ Wetlands, forests, grasslands, and other natural systems provide billions of dollars worth of services that most of us take for granted.
20.___ The term "debt-for-nature swap" refers to the ability of developing nations to take on foreign debt in order to protect species in foreign countries.

True/False Key:

1. T 2. F 3. T 4. F 5. T 6. T 7. T 8. F 9. T 10. F 11. T 12. F 13. T 14. F 15. T 16. T 17. F 18. T 19. T 20. F

Fill-in-the-Blank Questions

1. A _____ zone is a region in which limited human activity is allowed that surrounds a protected area.
2. Land set aside for use by native people on a sustainable basis is called an _____ reserve.
3. _____ house many endangered species and operate breeding programs for some species.
4. In 1973, the U.S. Congress passed the _____ Species Act which requires identification and listing of endangered and threatened species in the United States.
5. The industry that caters to the demands of bird watchers, wildlife lovers who like to travel is called _____.
6. A species that is in imminent danger of becoming extinct is called an _____ species.
7. _____ hunting is the killing of animals to provide food for indigenous people such as those that live in the tropical rain forests.
8. Peregrine falcon populations were severely damaged by the use of the persistent pesticide ___.
9. Illegal hunting is called _____.
10. The gopher tortoise and fig trees would be considered to be _____ species.
11. _____ alteration is the number one cause of species extinction.
12. _____ hunting and harvesting of wild species represent the second largest threat to the world's animal species.
13. Pollution by the heavy metal _____ has threatened wildlife at the Kesterson National Wildlife Refuge in California.

14. Global warming caused by _____ _____ pollution from the burning of fossil fuels may be responsible for the massive die-off of the world's coral reefs.
15. A _____ species is one that is still abundant in its natural range but, because its numbers are declining, is likely to become extinct in time.
16. _____ extinction is the loss of species largely as a result of human activities.
17. The sea _____ is a keystone species that eats sea urchins, abalone, crabs and molluscs.
18. Chemical _____ , sprayed on farms to control insect pests, and predator control programs have had a profound impact on native species.
19. The English sparrow was introduced into the United States in _____ .
20. Arguments for protecting endangered species and preserving biodiversity can be made on _____ and ethical grounds.

Fill-in Key:

1. buffer
2. extractive
3. Zoos
4. endangered
5. ecotourism
6. endangered
7. subsistence
8. DDT
9. poaching
10. keystone
11. habitat
12. commercial
13. selenium
14. carbon dioxide
15. threatened
16. accelerated
17. otter
18. pesticides
19. 1850
20. Utilitarian

Multiple Choice Questions

1. An endangered species is one that is in imminent threat of:
   a. population reduction.
   b. population explosion.
   * c. extinction.
   d. having future problems.
   e. being listed by the government

2. Today, approximately _____ species are threatened with extinction.
   a. 5,000
   b. 10,000
   c. 15,000
   d. 20,000
   * e. 25,000

3. Which of the following is a factor contributing to the loss of species?
   a. habitat alteration
   b. commercial harvesting
   c. introduction of alien species
   d. pest and predator control
   * e. all of the above

4. The killing of animals for food by indigenous people is called _____ hunting.
   a. sport
   * b. subsistence   d. market
   c. commercial   e. Tribal

5. Which of the following responses is a biological factor that could contribute to extinction?
   a. number of offspring
   b. size of the animal's range
   c. tolerance for humans
   d. their degree of specialization
   * e. all of the above

6. The Gopher tortoise share their burrows with many other organisms; it is a/an:
   a. predator
   b. important prey item
   * c. keystone species
   d. producer
   e. competitor with other species

7. Introduction of alien species is an ecological concern because they can cause extinction of native species by:
   a. migration
   b. parasitism
   * c. competition
   d. stimulation reproduction
   e. invasion

8. A good example of an alien bird species that has spread quickly throughout the North American continent is the
   a. robin
   b. crow
   * c. starling
   d. mallard duck
   e. goldfinch

9. These areas are particularly vulnerable to invasion by alien species:
   a. deserts
   b. lakes
   * c. islands
   d. marshes
   e. grasslands

10. Which of the following is a reason that biodiversity should be protected?
    a. aesthetics and economics
    b. food, pharmaceuticals
    c. protecting free services
    d. ethics
    * e. all of the above

11. About ___% of all prescription and nonprescription drugs are made with chemicals derived from wild plants
    a. 10
    b. 20
    c. 30
    d. 40
    * e. 50

108

12. Studies have shown that three-fourths of the birds and mammals of the Amazonian rain forest rely on the _____.
    a. Gopher tortoise
    b. rubber tree
*   c. fig tree
    d. balsa tree
    e. mahogany tree

13. This keystone predator in the marine ecosystem of the U.S. Pacific feeds on sea urchins, abalone, crabs, and mollusks that inhabit kelp beds:
*   a. sea otter
    b. blue whale
    c. barracuda
    d. sea lion
    e. walrus

14. This pollutant caused eggshell thinning and devastated many predatory bird populations:
    a. mercury
    b. lead
    c. calcium
    d. malathion
*   e. DDT

15. The idea that nations who have foreign debt might set aside important land for protection in exchange for forgiveness of their loans is called:
*   a. debt-for-nature swap
    b. creative financing
    c. sustainable development
    d. debt relief
    e. ecological financing

16. Areas that permit wildlife to move between one protected area and another are called:
    a. ecological islands
    b. buffer zones
*   c. wildlife corridors
    d. strip development
    e. conservation easements

17. Protecting free services applies to which of the following activities?
    a. road construction
*   b. insect control by birds
    c. pesticide application
    d. water treatment
    e. air pollution control

18. The Endangered Species Act requires the _____ to designate and list endangered and threatened species in the U.S..
    a. Department of Defense
*   b. U.S. Fish and Wildlife Service
    c. Forest Service
    d. Bureau of Land Management
    e. Department of Agriculture

19. This organism was endangered and caused a temporary halt of construction of the Tennessee Valley Authority's Tellico Dam.
    a. grass carp
    b. brook trout
*   c. snail darter
    d. emerald shiner
    e. Colorado chub

20. This marine mammal has been commercially hunted since the 1700s causing many species to decline to the brink of extinction:
    a. sea otters
    b. manatees
*   c. whales
    d. sea lions
    e. walruses

21. Pollution of the water in Kesterson National Wildlife Refuge in California with this metal was responsible for disastrous effects on a wide variety of wildlife:
    a. lead
    b. mercury
    c. magnesium
*   d. selenium
    e. aluminum

22. Isolated patches of protected habitat are called:
    a. buffer zones
    b. wildlife corridors
    c. reserves
    d. extractive reserves
    * e. ecological islands

23. Land set aside for native people to use on a sustainable basis is called a/an:
    * a. extractive reserve
    b. buffer zone
    c. refuge
    d. reserve
    e. wilderness area

24. A region in which limited human activity is allowed surrounding a protected area is called a:
    a. extractive reserve
    * b. buffer zone
    c. reserve
    d. roadless area
    e. wilderness area

25. The world's first extractive reserve was in:
    a. Niger
    b. Kenya
    c. Canada
    * d. Brazil
    e. Chile

26. Ecotourism is usually perceived as a sustainable activity. Which of the following responses is an environmental concern regarding this activity?
    * a. Tour vehicles can cause environmental damage/
    b. Tourists support local economies.
    c. Tourists promote keeping the areas they visit undeveloped.
    d. Tourists learn about the local people and customs.
    e. Local peoples realize that a healthy environment pays.

27. This now extinct bird was exterminated by widespread commercial hunting and habitat destruction in the U.S.:
    a. ringneck pheasant
    b. peregrine falcon
    * c. passenger pigeon
    d. Egyptian goose
    e. black duck

28. In the Pacific Northwest the building of hydroelectric dams and habitat destruction by logging is threatening many populations of:
    a. carp
    b. catfish
    * c. salmon
    d. halibut
    e. cod

29. This giant California bird has been brought to the brink of extinction by DDT, habitat destruction, and fire suppression:
    a. golden eagle
    b. Swainson's hawk
    * c. condor
    d. peregrine falcon
    e. bald eagle

30. Recently, this species has halted logging in old-growth forests in Oregon and Washington:
    a. sockeye salmon
    b. grizzly bear
    c. Uinta ground squirrel
    * d. spotted owl
    e. bull trout

31. Zoos are important resources because they provide:
    * a. place for breeding of endangered animals.
    b. low cost alternative to "real" habitats.
    c. nice place to see animals.
    d. refuge for animals from the circus.
    e. place to develop new habitats.

32 From the oceans millions of tons of fish are harvested worth _____ billion a year.
    a. $30
    b. $40
    c. $50
    d. $60
*   e. $70

33. Vegetated land helps to:
    a. replenish groundwater supplies
    b. reduce flooding
    c. control erosion
    d. maintain local climate
*   e. all of the above

34. Mangrove swamps provide an important free service to society, they help:
    a. purify water.
    b. fish spawning.
    c. flood control.
    d. provide timber
*   e. all of the above

35. The primary criteria used to decide if species are listed on the endangered list is:
    a. public opinion and press coverage.
    b. the visibility of the animal and its size.
    c. how cute the animal is.
    d. where the animal is located and the political power of this region.
*   e. population size and rate of population increase.

36. Decimation of important predator populations is ecologically dangerous because predators:
    a. have high rates of population growth.
    b. will respond in unknown ways to low population.
    c. are harmful animals and can kill.
    d. they can never be exterminated.
*   e. are the primary factors controlling the size of many prey populations.

37. This country contains 10% of the world's species:
    a. Brazil
    b. India
    c. Costa Rica
    d. Australia
*   e. China

38. Which of the following people proposed the *debt for nature swap*?
*   a. Dr. Thomas Lovejoy
    b. Dr. Paul Ehrlich
    c. Norman Myers
    d. Bob Davison
    e. E.O. Wilson

39. The rate of natural extinction is about one species every _____ years.
    a. 10
    b. 100
*   c. 1000
    d. 10,000
    e. 100,000

40. Estimates suggest that our current rate for loss of species is as high as _____ species per day.
    a. 5 to 10
    b. 10 to 20
    c. 30 to 40
*   d. 40 to 100
    e. 100 to 200

13

Grasslands, Forests, and Wilderness: Sustainable Management Strategies

Chapter Outline

The Tragedy of the Commons

Rangelands and Range Management: Protecting the World's Grasslands
An Introduction to Rangeland Ecology
The Condition of the World's Rangeland
Rangeland Management: A Sustainable Approach
Revamping Government Policies
Sustainable Livestock Production

Forests and Forest Management
Status Report on the World's Forests
Root Causes of Global Deforestation
An Introduction to Forest Harvest and Management
Creating a Sustainable System of Forestry

Wilderness and Wilderness Management
Why Save Wilderness?
Preservation: The Wilderness Act
Sustainable Wilderness Management

Key Terms

communal resources	commons	rangelands
grasslands	basal zone	metabolic reserve
range management	deferred grazing	feedlots
deforestation	clear-cutting	surface runoff
sublimation	mychorrizal fungi	selective cutting
strip cutting	shelter-wood cutting	sustainable forestry
second growth forests	crown fire	prescribed burns
primary forests	wilderness	primitive areas
wilderness areas		

Objectives

1. Describe the "tragedy of the commons" and its message for management of public lands.
2. Define "rangeland", "basal zone" and "metabolic reserve".
3. List the principles of sustainable rangeland management.
4. Discuss the root causes of deforestation.
5. Summarize how government policies concerning forestry have contributed to deforestation.
6. List the methods of harvesting forests.
7. Discuss the four measures required to create a sustainable system of wood production.
8. Define the following terms: "sustainable forestry", "prescribed fires", "crown fires", and "primary forests".
9. Summarize the legislation affecting "wilderness areas" in the United States.
10. List what you can you personally to protect rangeland, forests and wilderness.

Lecture Outline

I. The Tragedy of the Commons
   A. A commons is any public commodity or resource.
   B. The tragedy of the commons is that individuals are compelled to abuse a commons in pursuit of personal gain.
   C. Privately-owned land is often similarly abused; again, the lure of short-term profits compels some to ignore long-term damage.

II. Rangelands and Range Management: Protecting the World's Grasslands - Rangelands and their produces are potentially renewable resources, if properly managed; if not, they can be ruined.
   A. An Introduction to Rangeland Ecology
      1. Grasses form the base of the food chain on rangelands. Grasses are adapted to periodic fire, drought, and grazing so long as care is taken to protect the metabolic reserve of the plant.
   B. Rangeland Deterioration - Most rangeland in the U.S., both federal and private, has deteriorated due to overuse and mismanagement.
   C. Rangeland Management: A Sustainable Approach
      1. Controlling Livestock Numbers - Rangeland and pasture use must be adjusted according to the carrying capacity of the land, which varies from one year to the next with the weather.
      2. Deferred Grazing - Cattle can be withheld from rangeland to permit grasses to mature and produce seeds. This method enhances rangeland and may increase the carrying capacity in the long run.
      3. Controlling the Distribution of Livestock - Careful distribution of water sources and salt lick can help promote a more uniform use of rangeland and protect some area from serious degradation.
      4. Restoring and Improving the Quality of Rangeland - Restoration of grasslands is an essential element of building a sustainable system of production. Periodic burning can also be used to increase rangeland productivity.

5. Revamping Government Policies -
   a. The Public Rangelands Improvement Act of 1978 was a first step towards reducing damage and developing sustainable rangeland management practices in the U.S.
   b. The Public Rangelands Improvement Act requires the BLM and the U.S. Forest Service to reduce grazing on public land where damage is evident.
D. Sustainable Livestock Production - In many countries, livestock are raised in pens and fed grains that could be used to feed large numbers of people. In many countries meat production may have to be reduced to accommodate the food demands of a growing population.

III. Forests and Forest Management - Forests provide tremendous social, economic, and environmental benefits. Despite their importance to us, forests are in a state of decline.
A. Status Report on the World's Forests
   1. Worldwide, about one-third of the world's forests have been cut.
   2. Deforestation continues today with tremendous losses in some areas: 32 million hectares (78 million acres) of tropical rain forests are leveled each year. At the current rate of harvest China will lose all of its commercial forests within 10 years.
   3. The deforested land is too often converted to other uses, mostly farming. In developing countries for every 10 trees cut only 1 tree is replanted
   4. Deforestation not only destroys habitat for many species but decreases sustainable fuel supplies needed for cooking and home heating in some areas
B. Root Causes of Global Deforestation - Many factors contribute to the continuing pattern of deforestation including:
   1. Population Growth - Today 25%-60% of the annual tropical deforestation is attributed to humans in search of sustenance.
   2. Poverty - A natural response to poverty is movement into forested areas to eke out a living.
   3. Inequitable Land Ownership - In developing nations prime agricultural land may be owned by a small number of people. Poor rural peasants often seek the forested areas for farming and clear the land.
   4. Unsustainable Government Policies -
      a. Governments have typically sold timber below market value to logging companies.
      b. Many economic policies encourage unsustainable exploitation of the forests.
         i. Many developing countries restrict the export of raw wood by international companies in the desire to keep jobs in their own countries. Many of the lumber mills in these countries are inefficient and use 50% more logs than the industry standard thereby encouraging more deforestation.
         ii. Short-term contracts are encouraged by companies even though a longer time is necessary for the forest to recover after harvesting.
         iii. Heavy foreign debt in some nations encourages excessive timber harvest to pay back loans.
         iv. Government tax policies often encourage deforestation. In Brazil, the government once offered huge income tax credits to investors in cattle ranches, once a leading cause of deforestation in the Amazon. The Canadian government has encouraged deforestation with little controls for environmental quality. Currently, this is threatening forests in

114

western Canada and has endangered important salmon fisheries and grizzly bear habitat.

C. An Introduction to Forest Harvest and Management - Trees are harvested primarily in four ways:
1. Clear-cutting - Clear-cutting is the complete removal of trees from a tract of land. While sometimes benefiting certain game species, this practice more often damages habitat by accelerating erosion, surface runoff, and sedimentation.
2. Strip-Cutting – Clear-cutting can be carried out on a smaller scale to minimize visual and environmental impacts.
3. Selective Timber Cutting - Selective cutting involves the removal of select trees from an otherwise more-or-less intact forest. Properly employed, selective harvesting is the least environmentally method of tree harvest, though improper application may damage a forest immediately and over the long term.
4. Shelter-wood Cutting - This is a type of selective harvesting where poor-quality tress are first removed from mixed timber stands leaving the healthiest trees intact. Once seedling from the existing stand become established loggers remove a portion of the commercially valuable mature trees leaving enough in place to provide shade for the seedlings. When the seedling become saplings the remaining mature trees are harvested.

D. Creating a Sustainable System of Forestry - Four general measures are required to create a sustainable system of forestry:
1. Reducing Demand
   a. Reducing the growth of the human population would reduce pressure on the world's forests.
   b. Wood can be saved by using thinner saw blades, improved machinery to grind up logs, and other innovative technologies.
   c. Reducing consumption of wood products by increasing construction/design efficiencies.
   d. Recycling or wood and paper products.
   e. The use of alternative building materials (composite beams, straw bales, etc.).
2. Managing Forests and Tree Farms Sustainably
   a. Clear-cutting should be more carefully regulated and reseeding programs carefully monitored.
   b. Forests should be protected from natural hazards.
   c. Prescribed burns and thinning techniques should be used to create a healthier, more diverse forest.
3. Saving Primary Forests/Creating Forest Preserves
   a. Saving uncut or primary forests helps protect biodiversity but also protects nearby harvested forests from outbreaks of pests.
4. Restoring Forestland - Restoration is a key principle of sustainability and vital to continued timber production.

IV. Wilderness and Wilderness Management - Large tracts of wilderness exist today but pressure is mounting to develop many of these pristine areas.
A. Why Save Wilderness? Wilderness holds a variety of values for us, including aesthetic, recreational, economic, and spiritual benefits.
B. Preservation: The Wilderness Act - Passed in 1964, the Wilderness Act seeks to preserve certain lands in perpetual wilderness.
C. Sustainable Wilderness Management
1. Many popular wilderness areas are pressured by high levels of human use.

2. Crowding and the environmental degradation from overuse can be reduce or eliminated by some of the following suggestions:
   a. Campers can be educated on ways to lessen their impact.
   b. Access to overused areas can be restricted.
   c. Permits can be issued to control the number of users.
   d. Campsites can be designated to prevent additional damage.
   e. Information can be disseminated about areas that receive less use.
   f. Trails can be improved to promote underutilized areas.
3. Globally, interest in wilderness protection is growing. Ecotourism can provide a powerful incentive to protect wilderness in less developed nations.

Suggestions for Presenting the Chapter

- If you have a National Forest nearby a visit to the office and tour is an excellent exercise.
- If you cannot visit a forest (state or national) then you can obtain materials by contacting you state agency in charge or the Forest Service. Free publications including visitor maps and management plans are available at no cost. by contacting the U.S. Forest Service.
- A fieldtrip to a National Wildlife Refuge or Wilderness Area is another excellent choice to illuminate the management concepts discussed in the text.
- Many western states have grassland areas managed by the Bureau of Land Management or Forest Service. These are excellent areas to visit and maps/information are readily available. If you cannot visit one of these areas contact the BLM or Forest Service for details information or visit their Web sites.
- Recommended web sites:

  Bureau of Land Management/Department of the Interior: http://www.blm.gov/nhp/index.htm
  National Grasslands: http://www.fs.fed.us/grasslands/
  U.S. Forest Service: http://www.fs.fed.us/
  Sierra Club: http://www.sierraclub.org/
  The Wilderness Society: http://www.wilderness.org/
  The National Audubon Society: http://www.audubon.org/

True/False Questions

1. ___ The grasslands on which many of the world's livestock depend are know as rangelands.
2. ___ The National Park Service manages most of the federally owned rangeland.
3. ___ Grasses grow from the basal zone, the lowest portion of the leaf.
4. ___ Grasses are poorly adapted to grazing pressure, as a result, cattle should be grazed in timbered areas.
5. ___ Grasses can withstand drought because their roots extend deep into underground water supplies.
6. ___ Restoration of grasslands is an essential element of building a sustainable system of food production.
7. ___ Livestock can be raised in confined areas called feedlots.
8. ___ The Public Rangelands Improvement Act was passed by Congress in 1878.
9. ___ Forests cover about three-fourths of the Earth's land surface.

10.___ Deforestation results from frontierism, population growth, poverty, and inequitable land ownership.
11.___ Clear-cutting is the most environmentally friendly harvest method.
12.___ Clear-cutting is one of the fastest and cheapest methods of harvesting trees.
13.___ Strip-cutting is more damaging to the forest than clear-cutting.
14.___ The Forest Service deliberately sets hundreds of fires each year to remove underbrush and litter.
15.___ Wilderness is land not significantly altered by human activities.
16.___ The earliest efforts at wilderness preservation in the United States started in the 1960s.
17.___ The Federal Land Policy and Management Act (1976) required the Forest Service to submit recommendations on the wilderness suitability of its lands.
18.___ As a result of the Wilderness Act, the Forest Service's primitive areas were renamed wilderness areas.
19.___ John Muir was the founder of the Sierra Club.
20.___ The Spotted Owl is threatened as a result of clear-cutting in the Appalachian Mountains.

True/False Key:

1. 1. T 2. F 3. T 4. F 5. T 6. T 7. T 8. F 9. F 10. T 11. F 12. T 13. F 14. T 15. T 16. F 17. F 18. T 19. T 20. F

Fill-in-the-Blank Questions

1. _____ cutting is the removal of a limited number of trees form a forest.
2. _____ are the grasslands that are used by grazing animals.
3. _____ form the base of the food chain on rangelands.
4. Grasses can withstand _____ because their roots extend deep into underground water supplies.
5. The Bureau of Land _____ manages most of the federally owned cropland.
6. Withholding cattle from rangeland and allowing the grasses to mature and produce seed is called _____ grazing.
7. The U.S. Congress passed the Public Rangelands Improvement Act in _____.
8. Deforestation results from many factors including _____, population growth, poverty and inequitable land ownership.
9. Clear-cutting is a standard method of harvest that is used primarily for _____ that grow in large stands containing few tree species.
10. Approximately ___% of the nutrients of a rainforest are in biomass and only 5% are in the soil.
11. In ____, Brazil established the first extractive reserve in the world.
12. _____ cutting is a kind of selective harvest method where poor-quality trees are removed first from mixed timber stands, leaving the healthiest trees intact.
13. The remaining uncut forests are called _____ forests.

14. _____ is land not significantly altered by human activities.
15. In _____, Congress passed the Wilderness Act establishing the National Wilderness Preservation System.
16. Better management of existing forests, including tree _____ and prescribe burns, helps to create a more diverse and healthy forest.
17. President _____ _____ established the U.S. Forest Service in 1905.
18. According to the Forest Service, ___% of all forest fires are caused by human beings.
19. _____ burns are used by the Forest Service to remove underbrush and litter.
20. The conversion of snow to water vapor is called _____.

Fill-in Key:

1. selective
2. rangelands
3. grasses
4. drought
5. Management
6. Deferred
7. 1978
8. frontierism
9. softwoods (conifers)
10. 95
11. 1988
12. shelter-wood
13. primary
14. wilderness
15. 1964
16. thinning
17. Theodore Roosevelt
18. 85
19. prescribed
20. Sublimation

Multiple Choice Questions

1. The primary message of the "tragedy of the commons" is :
   a. Private land is always well managed.
   b. Public land is always well managed.
   c. Public and private lands are well managed.
   * d. The possibility for abuse of public held land is real.
   e. Pubic land should become privatized.

2. The grasslands on which many of the world's livestock depend are known as:
   a. fields.
   b. desert.
   c. steppes.
   * d. rangelands.
   e. Feedlots

3. In the United States _____ million acres of grassland have been turned into desert in the last 200 years.
   a. 50
   b. 100
   * c. 225
   d. 550
   e. 1050

4. According to the Bureau of Land Management, _____% of the public rangeland is in fair-to-poor shape.
   a. 12
   b. 32
   c. 43
   d. 55
   * e. 62

5. The United Nations Environmental Programme estimates that nearly _____% of the world's rangeland is at least moderately desertified.
   a. 25
   b. 50
   * c. 75
   d. 85
   e. 100

6. Grasses can be grazed without damage because they grow:
   a. when they are fertilized.
   b. during the summer months.
   * c. from the basal zone.
   d. only once a year.
   e. without water or nutrients.

7. The metabolic reserve is the:
   a. amount of energy all animals have.
   b. the amount of energy left after exercise.
   * c. the bottom portion of the grass plant.
   d. the top portion of the grass plant.
   e. only part that can be eaten by cattle.

8. The Public Rangelands Improvement Act was passed by Congress in _____.
   a. 1948
   b. 1958
   c. 1968
   * d. 1978
   e. 1988

9. Feedlots are environmentally unsound because they use:
   a. too much hay .
   * b. enormous quantities of grain.
   c. excessive amounts of water.
   d. new nutrient recycling techniques.
   e. too much confinement space.

10. Forests cover about _____ of the Earth's land surface.
    a. one-tenth
    b. one-sixth
    c. one-fifth
    d. one-fourth
    * e. one-third

11. Primary products of the forest are:
    * a. logs
    b. paper
    c. lumber
    d. furniture
    e. turpentine

12. The most damaging threat to good rangeland is:
    a. soil erosion
    b. grass fungus
    c. grasshoppers
    d. sheep
    * e. overgrazing

13. Which of the following is a basic technique used in range management?
    a. controlling livestock numbers
    b. deferred grazing
    c. controlling distribution of livestock
    d. restoration
    * e. all of the above

14. Withholding cattle grazing on rangeland is called:
    a. rotation
    * b. deferred grazing
    c. restoration
    d. grazing allotment
    e. feedlot time

15. The United States has about _____ million acres of forestland.
    a. 340
    b. 440
    c. 540
    d. 640
    * e. 740

16. Only about __% of the world's forest land is under any kind of management.
    a. 3
    * b. 13
    c. 23
    d. 33
    e. 43

17. Which of the following things should you do to protect rangelands?
    a. Learn more about them.
    b. Contact rangeland expert's in your state.
    c. Reduce your meat consumption.
    d. Write your Congressional representative to support measures to enhance rangeland conditions.
    * e. all of the above

18. Secondary products of the forest are:
    a. logs
    * b. paper
    c. furniture
    d. turpentine
    e. wood oils

19. Tertiary products of the forest are:
    a. logs
    b. paper
    c. lumber
    d. particle board
    * e. turpentine

20. Loggers remove all the trees in a 40-200 acre plot in this forestry technique:
    a. shelter-wood cutting.
    b. selective cutting.
    c. strip-cutting.
    * d. clear-cutting
    e. thinning

21. Which of the following responses is an environmental problem caused by clear-cutting?
    a. sublimation
    b. surface runoff
    c. destroys nutrient-cycling bacteria
    d. fragments wildlife habitat
    * e. all of the above

22. The conversion of snow to water vapor is:
    a. snowmelt
    b. evaporation
    c. supercooling
    * d. sublimation
    e. superheating

23. This forestry technique harvests trees in a narrow strip that blend with the terrain:
    a. thinning
    b. selective-cutting
    c. clear-cutting
    * d. strip-cutting
    e. shelter-wood cutting

24. Which of the following responses is a factor causing global deforestation?
    a. frontierism
    b. lack of knowledge about the importance of forests
    c. population growth
    d. poverty
*   e. all of the above

25. Which of the following activities is an unsustainable government policy contributing to deforestation?
    a. below cost timber sales
    b. short-term contracts for timber sale
    c. foreign debt in developing countries
    d. economic policies that encourage inefficient use
*   e. all of the above

26. Which of the following is a general measure to create a sustainable system of wood production?
    a. Reduce demand for wood and wood products.
    b. Use sustainable management practices.
    c. Establish forest reserves.
    d. Restore existing forest lands.
*   e. all of the above

27. What can be done to reduce demand for wood and wood products?
    a. Use thinner saw blades.
    b. Build smaller homes.
    c. Recycle paper and wood.
    d. Use alternative building materials.
*   e. all of the above

28. Which of the following Presidents established the Forest Service?
    a. Grover Cleveland
    b. Ulysses Grant
    c. James Garfield
*   d. Theodore Roosevelt
    e. Harry Truman

29. This forestry technique harvests only mature trees of a desired species:
    a. thinning
*   b. selective-cutting
    c. clear-cutting
    d. strip-cutting
    e. shelter-wood cutting

30. This particular way to harvest forests may be economically competitive with clear-cutting in second growth forests:
    a.. thinning
    b. selective-cutting
    c. shelter-wood cutting
    d. strip-cutting
*   e. b and c only

31. About ___% of the world's wilderness lands are legally protected.
    a. 10
*   b. 20
    c. 30
    d. 40
    e. 50

32. The Wilderness Act of 1964 established the:
    a. primitive area
    b. Fish and Wildlife Service
*   c. National Wilderness Preservation System
    d. Sierra Club
    e. Federal Land Policy and Management Act

33. The state with the largest designated wilderness acreage is:
    a. Wyoming
    b. Montana
*   c. Alaska
    d. Idaho
    e. Colorado

34. The major problem with the original Wilderness Act is that it did not provide for wilderness designation on:
    a. state lands.
    b. Indian lands.
*   c. Bureau of Land Management lands.
    d. Department of Energy lands.
    e. Department of Defense lands.

121

35. Fire accounts for ___% of U.S. forest destruction yearly.
    a. 7
*   b. 17
    c. 27
    d. 37
    e. 47

36. About 85% of all forest fires are caused by:
    a. prescribed burns.
*   b. humans.
    c. lightning.
    d. war.
    e. volcanism.

37. Land not significantly altered by human activities is called a:
    a. rangeland
    b. preserve
    c. refuge
*   d. wilderness
    e. park

38. The first head of the U.S. Forest Service was:
    a. Theodore Roosevelt
    b. John Muir
    c. Bob Marshall
*   d. Gifford Pinchot
    e. Aldo Leopold

39. The Forest Service deliberately sets hundreds of fires each year these are:
    a. crown fires
    b. back fires
*   c. prescribed fires
    d. fires in clear-cuts
    e. wilderness fires

40. Who was founder of the Sierra Club?
    a. Rachel Carson
    b. Theodore Roosevelt
    c. Gifford Pinchot
*   d. John Muir
    e. Ralph Waldo Emerson

14

Water Resources: Preserving Our Liquid Assets and Protecting Aquatic Ecosystems

Chapter Outline

The Hydrological Cycle

Water Shortages
Where is Water Used and Where Does It Come From?
Drought and Water Shortages
Impacts of the Water Supply System
Impacts of Excessive Groundwater Withdrawals
Impacts of Dams and Reservoirs
Creating a Sustainable Water Supply System

Flooding: Problems and Solutions
Causes of Flooding
Controlling Floods

Wetlands, Estuaries, Coastlines, and Rivers
Wetlands
Estuaries
Barrier Islands
Coastal Beaches
Wild and Scenic Rivers

Key Terms

hydrological cycle	evaporation	transpiration
evapotranspiration	absolute humidity	relative humidity
condensation nuclei	precipitation	surface waters
ground water	stable runoff	drought
water table	saltwater intrusion	subsidence

sinkholes	aquifer recharge zones	watersheds
wetlands	xeriscape	gray water
black water	flooding	flood plains
levees	streambed channelization	watershed management
estuaries	barrier islands	coastal beaches
jetties	beach drift	longshore currents
longshore drift	wild rivers	scenic rivers
recreational rivers		

Objectives

1. Draw a sketch of the hydrological cycle showing its major processes and components.
2. Define the following terms: "evaporation", "transpiration", "evapotranspiration", "absolute humidity", and "relative humidity".
3. Summarize the major sources of human use of water in the United States and globally.
4. Define the following terms: "surface waters", "ground water", "stable runoff", and "drought".
5. Discuss the potential impacts of excessive groundwater withdrawal.
6. Summarize the impacts of dams and reservoirs on the environment.
7. Discuss the ways the present system of water supply and use can be made more sustainable.
8. Summarize the major causes of flooding and how watershed management can reduce flooding.
9. List the major aquatic life zones associated with surface waters.
10. Discuss important legislation affecting aquatic habitats discussed in this chapter.
11. Discuss why wetlands are more than just important areas of habitat for aquatic plants and animals.
12. Define the terms: "wild rivers", "scenic rivers" and "recreational rivers".
13. Discuss how you can personally conserve water.

Lecture Outline

I. The Hydrological Cycle - Water is a renewable resource purified and distributed in the hydrological cycle.
   A. The hydrological or global water cycle consists of two phases, evaporation and precipitation.
   B. The cycle is driven by winds generated by solar energy.

II. Water Shortages - Water shortages are a function of uneven distribution, excessive demand and inefficient use.
   A. Where Is Water Used, and Where Does It Come From?
      1. Globally, agriculture and industry are the major users of water.
      2. Most water comes from surface water supplies: rivers, streams, and lakes.
      3. Withdrawal of water in excess of the *stable runoff* typically results in severe water shortages and hardships for people and many species.
   B. Drought and Water Shortages
      1. While droughts appear to be natural causes of water shortages, human activities may increase their frequency and severity.

2. Droughts reduce water supplies and create significant social, economic, and environmental problems.

C. Impacts of the Water Supply System
1. Water is withdrawn from groundwater and surface water sources, but very little is returned directly to its source.
2. Water that is returned to its source is often laden with pollutants.

D. Impacts of Excessive Groundwater Withdrawals
1. Groundwater Overdraft - Groundwater overdraft is the removal of groundwater at a rate faster than it is replaced. Groundwater overdraft is responsible for several important problems:
   a. Saltwater Intrusion - Saltwater intrusion is the movement of salt water from marine aquifers into freshwater aquifers. Salt water is then contaminating potable water supplies.
   b. Subsidence - Removal of groundwater can cause soil compaction and sinks or sinkholes. These can affect large areas of land making them unsuitable for human use.

E. Impacts of Dams and Reservoirs
1. Dams have many positive benefits such as power generation, flood control and water storage.
2. Dams can have a detrimental impact on fish and wildlife. They flood streams and can eliminate fish species that live and reproduce there. They also flood terrestrial habitats affecting large numbers of other species.
3. Dams can impact human activities and economies negatively. New dams can threaten recreational rivers and surrounding towns and their economies. Dams built in developing countries often flood valuable farmland and can encourage new epidemics of disease.
4. Dams can have downstream impacts on rivers, often changing the environmental conditions severely.
5. Water diversions as a result of dam projects can affect water quality and dewater downstream areas.

F. Creating a Sustainable Water Supply System
1. Water Conservation - traditional approaches to water problems have focused on supply: damming rivers and streams and pumping water from lakes and aquifers. A more sustainable approach is to conserve on the use side.
   a. Agriculture and industry are the biggest users and efforts to conserve in these sectors should be given the highest priority.
   b. Water projects are often subsidized by the government providing no incentives for conservation. Reducing the amount of irrigation water by improving efficiency or switching to crops that do not require irrigation are several strategies that could be employed.
   c. Sustainable farming practices reduce the demand for watering and irrigation.
2. Water Recycling - Water can be purified and reused in industry, on farms, and even in our homes. Water recycling reduces pressure on surface and groundwater supplies and reduces water pollution.
3. Restoring Watersheds and Wetlands to Protect and Enhance Our Water Supplies - Restoring watersheds reduces siltation in reservoirs and rivers and enhances groundwater recharge. Wetlands are the natural filters for surface water and restoration of these areas will reduce pollution of surface and ground water.
4. Population Stabilization - Stabilizing or reducing human population size can reduce the demand on the water system.
5. Changing Government Policies to Create a Sustainable Water Supply System

a. Government policies should be revamped to require those who benefit from water projects to pay the full cost. The cost of a water project should include any environmental costs should problems arise as a result of the project.

b. Governments should establish building codes that include requirements for water conserving devices.

c. Alternative landscaping method requiring less water should be subsidized and encouraged (xeriscape).

d. Gray water systems should be installed to reduce water use. Building codes can require these be installed in new construction.

e. Government can control water prices and adjust water rates by time of day and encourage conservation.

f. Education can play an important function in reducing water use.

III. Flooding: Problems and Solutions - Despite widespread water shortages, flooding is a serious problem in many parts of the world.

A. Causes of Flooding - Human activities and land use patterns combine with heavy precipitation and other natural factors to produce floods.

B. Controlling Flooding - Structural flood-control devices and watershed management are two different approaches to controlling flooding.

C. Streambed channelization is an environmentally destructive technique. The construction of floodwall not only eliminates many important wetland areas but prevents normal flooding and enrichment of the flood plain ecosystem. In addition, channelization has been shown to exacerbate the effects of flooding in downstream areas.

IV. Wetlands, Estuaries, Coastlines, and Rivers - Surface waters are under assault. The destruction of these systems affects available water supplies but also eliminates important habitat for fish and other organisms.

A. Wetlands - These are flooded lands lying either inland or in coastal zones.

1. The Hidden Value of Wetlands - Wetlands function as valuable habitats and water-quality protectors.

2. Declining wetlands - Wetlands are destroyed in the U.S. and elsewhere for development and agriculture.

3. Protecting Wetlands - Various federal and state bills are designed to protect remaining wetlands in the U.S., though enforcement, and thus effectiveness, is low.

B. Estuaries - Estuaries, or river mouths and bays, are critical habitat for many commercial species and wildlife. Estuarine zones are valuable also as water purifiers.

1. Damaging This Important Zone - Pollution, sedimentation, inflow reduction, and dredging/filling projects all harm estuarine zones and their life forms.

2. Protecting the Estuarine Zone - Pollution and erosion control, along with water conservation, will help protect the estuarine zone.

C. Barrier Islands - Barrier islands form in response to wave and wind patterns offshore. Development and beach-erosion control programs threaten both barrier islands and coastal beaches.

D. Coastal Beaches - Like barrier islands, coastal beaches are eroded by longshore currents.

1. Protecting coastlines may require measures to ensure a steady stream of sediment from rivers and a hands-off policy toward building levees.

E. Wild and Scenic Rivers -

126

1. Rivers provide many benefits in addition to recreation. They are great reservoirs of diversity and should be respected as such. This means that interruption of in-stream flow is probably not a sustainable activity.
2. Many rivers or segments of rivers in the United States have been given protection against development by the Wild and Scenic Rivers Act. The wild or scenic river designation is often marked by controversy because so many interests vie for a river's benefits.
3. Dammed and diverted to water-hungry, often wasteful consumers, a river becomes a symbol of unsustainable systems design.

Suggestions for Presenting the Chapter

- A visit to a local lake, river, estuary or coastline to investigate the ecosystem is an excellent way to introduce your students to aquatic ecosystems.
- Instructors should stress that the health and sustainability of aquatic ecosystems is dependent on the sustainability of human activities using land as well as water.
- A trip to your local water plant and/or a waste water/sewage treatment plant is an excellent activity. Students should know where and how their potable water is obtained and how wastewater is treated before returning to surface waters.
- Instructors should stress that human development directly influences our water resources. Pollution of ground and surface waters, water supply problems, and flooding are just some of the main issues associated with human development.
- Recommended web sites:

  The Online Meterorology Guide: http://ww2010.atmos.uiuc.edu/(Gh)/guides/mtr/home.rxml
  National Weather Service Homepage/NOAA: http://www.nws.noaa.gov/
  American Rivers: http://www.amrivers.org/
  National Inventory of Dams/U.S. Army Corps of Engineers:
  http://crunch.tec.army.mil/nid/webpages/nid.cfm
  Wild and Scenic Rivers/National Park Service: http://www.nps.gov/rivers/

True/False Questions

1. ___ Water is part of a global recycling network known as the hydrological cycle.
2. ___ The evaporative loss of water from soil and leaves is called transpiration.
3. ___ The absolute humidity is the number of grams of water in a kilogram of air.
4. ___ If the relative humidity is 50% the air is said to be saturated.
5. ___ Water molecules attach onto various small suspended particles called condensation nuclei in the atmosphere.
6. ___ Precipitation returns water to lakes, rivers, oceans, and land.
7. ___ At any single moment 50% of the Earth's water is in the oceans.
8. ___ Most of the Earth's freshwater is locked up in the polar ice, glaciers and in deep aquifers.
9. ___ Globally agriculture and industry are the major users of water.
10. ___ Most of the water used by humans comes from groundwater sources.
11. ___ 66% of the precipitation reaching earth evaporates back into the atmosphere.
12. ___ Saltwater intrusion results from the movement of salt water from marine aquifers into freshwater aquifers.

13.___ A severe drought results in a decrease in stream flow and an increase in the water table.
14.___ Regions where rain and snowmelt replenish aquifers is the aquifer discharge zone.
15.___ The water from showers, sinks, and washing machines is called the gray water.
16.___ Levees are embankments constructed along the banks of rivers to hold floodwaters back.
17.___ Inland wetlands are wet or flooded regions along coastlines.
18.___ Estuaries are the mouths of rivers where salt water and freshwater mix.
19.___ The U.S. Congress passed the Wild and Scenic Rivers Act in 1988.
20.___ Scenic rivers are readily accessible by road and may have some dams along their course or development along their shores.

True/False Key:

1. T 2. F 3. T 4. F 5. T 6. T 7. F 8. T 9. T 10. F 11. T 12. T 13. F 14. F 15. T 16. T 17. F 18. T 19. F 20. F

Fill-in-the-Blank Questions

1. ____ rivers are rivers or sections of rivers that are relatively inaccessible and untamed.
2. The gradual movement of sand along a beach is called beach _____.
3. The Coastal Barrier Resources Act of 1982 was established to prohibit the expenditure of federal money for highway construction and other _____ on barrier islands.
4. The federal government can purchase wetlands and set them aside as part of the National Wildlife _____ System.
5. The federal Coastal Zone Management Act calls on states to develop plans to protect coastal _____.
6. _____ are perpetually or periodically flooded lands.
7. Streambed _____ is a kind of streamlining rivers produced by deepening and straightening river channels.
8. ____% of the water on Earth is in the oceans.
9. Removing the salts from seawater for drinking and irrigation is called _____.
10. Water from showers, sinks and washing machines is called the _____ water.
11. The soil compacts and sink when groundwater is withdrawn, a process cause _____.
12. A drought exists when rainfall is ___% below average for a period of 21 days or longer.
13. _____ runoff is a the amount of water flowing in rivers and streams that one can count on from year to year.
14. The evaporative loss of water from leaves is called _____.
15. The _____ humidity is the amount of moisture present in air compared with the amount it could hold if fully saturated, at any given temperature.
16. The loss of water from the soil and leaves is called _____.

128

17. If the relative humidity is 100% the air is said to be _____.
18. Clouds form when water molecules attach onto various small suspended particles in the air called _____ nuclei.
19. Together, coastal wetlands and estuaries make up the _____ zone.
20. Between 1940 and 1971, _____ miles of U.S. streams were channelized by the Army Corps of Engineers and the Soil Conservation Service.

Fill-in Key:

1. Scenic
2. drift
3. development
4. Refuge
5. wetlands
6. wetlands
7. channelization
8. 97
9. desalination
10. gray
11. subsidence
12. 70
13. Stable
14. transpiration
15. relative
16. evapotranspiration
17. saturated
18. condensation
19. estuarine zone
20. 22,000

Multiple Choice Questions

1. The hydrological cycle is driven by:
   a. wind movement.
   b. kinetic energy.
   * c. solar energy.
   d. potential energy.
   e. rotation of the earth.

2. The loss of water from the soil and leaves is:
   a. evaporation
   b. transpiration
   * c. evapotranspiration
   d. sublimation
   e. condensation

3. The loss of water from leaves is:
   a. evaporation
   * b. transpiration
   c. evapotranspiration
   d. sublimation
   e. condensation

4. The number of grams of water in a kilogram of air is the:
   a. relative humidity
   * b. absolute humidity
   c. vapor pressure
   d. partial pressure of water
   e. dew point

5. The amount of moisture present in the air compared with the amount in fully saturated air at a particular temperature is the:
    a. absolute humidity
    * b. relative humidity
    c. vapor pressure
    d. partial pressure
    e. dew point

6. If the relative humidity is 100% the air is said to be:
    a. condensed
    * b. saturated
    c. at dew point
    d. a cloud
    e. a condensation nuclei

7. Small particles in the air can serve as a surface for water to cling to in clouds, these particles are called:
    a. solute
    b. particulates
    * c. condensation nuclei
    d. smog
    e. salts

8. Precipitation falls to earth in which of the following forms?
    a. rain
    b. snow
    c. sleet
    d. hail
    * e. all of the above

9. How much of the water that is precipitated evaporates on the average?
    a. one-tenth
    b. one-eighth
    c. one-fourth
    d. one-third
    * e. two-thirds

10. How much of the earth's freshwater is locked up in ice, in glaciers, and in deep aquifers?
    a. 10%
    b. 30%
    c. 50%
    d. 80%
    * e. 99%

11. The single largest source of water use globally is:
    a. urban and residential use
    * b. agriculture
    c. industrial
    d. drinking water
    e. government

12. In the U.S., what fraction of the freshwater comes from surface water sources (lakes, streams, and rivers)?
    a. one-fourth
    b. one-third
    c. one-half
    * d . three-fourths
    e. all

13. The amount of water flowing in rivers and streams that one can count on from year to year is the:
    * a. stable runoff
    b. discharge
    c. discharge rate
    d. instream flow
    e. average flow

14. This condition exists when rainfall is 70% below average for a period of 21 days or longer:
    a. dry period
    * b. drought
    c. doldrums
    d. chinook
    e. continental low

15. The water table is the:
    a. surface of a lake.
    b. amount of available freshwater.
    * c. upper surface of the groundwater.
    d. upper surface of surface water.
    e. highest water in the area.

16. How much of the agricultural water applied to crops evaporates?
    a. 10%
    b. 20%
    c. 40%
    d. 60%
    * e. 80%

17. The removal of groundwater faster than it can be replenished is:
    a. intrusion
    b. depletion
    * c. groundwater overdraft
    d. subsidence
    e. unstable runoff

18. The movement of salt water from marine aquifers into freshwater supplies is:
    a. subsidence
    b. depletion
    c. groundwater overdraft
    * d. saltwater intrusion
    e. salinization

19. The process where sinkholes are formed as a result of water withdrawal from an aquifer is called:
    a. groundwater mining
    b. depletion
    * c. subsidence
    d. intrusion
    e. overdraft

20. Which of the following responses is an environmental impact caused by dams?
    a. Elimination of native fishes.
    b. Blockage of fish migration.
    c. Reduced nutrient flow to estuaries.
    d. Alter water quality in streams.
    * e. all of the above

21. This body of salt water in the former Soviet Union has been reduced by massive withdrawal of water for irrigation?
    a. Caspian Sea
    b. Black Sea
    * c. Aral Sea
    d. Lake Baikal
    e. Lake Odessa

22. Regions where rain and snowmelt replenish aquifers are called:
    a. streams
    b. wetlands
    * c. aquifer recharge zones
    d. discharge areas
    e. marshes

23. The landscaping technique using water-conserving species is called:
    * a. xeriscape
    b. naturalism
    c. natural restoration
    d. landscape architecture
    e. desertification

24. Water from showers and faucets is called:
    a. clear water
    b. black water
    * c. gray water
    d. brown water
    e. potable water

25. Water from toilets is called:
    a. clear water
    * b. black water
    c. gray water
    d. brown water
    e. potable water

26. Which of the following structures provides sustainable flood control?
    a. levees
    b. dams
    c. streambed channelization
    d. removing marshes
    * e. wetlands

27. When 75% of the water flows over the earth's surface, ___% percolates into the soil.
    a. 10
    * b. 25
    c. 50
    d. 75
    e. 90

28. One of the most promising and sustainable approaches to reduce flooding is:
    a. streambed channelization
    b. dams
    * c. watershed management
    d. levees
    e. removing wetlands

29. Chesapeake Bay on the eastern coast of the U.S. is a/an:
    a. lake
    b. freshwater marsh
  * c. estuary
    d. river system
    e. tidal flat

30. Mangrove swamps, salt marshes, bays, and lagoons are:
    a. tidal areas
    b. inland wetlands
  * c. coastal wetlands.
    d. estuaries
    e. freshwater marshes

31. Inland wetlands include which of the following areas?
    a. bogs.
    b. swamps
    c. freshwater marshes
    d. river overflow lands
  * e. all of the above

32. Today less than _____ of America's wetlands remain.
    a. one-eighth
    b. one-fourth
  * c. one-third
    d. one-half
    e. three-fourths

33 This aquatic habitat is found at the mouths of rivers and are important life zones:
    a. bayou
    b. swamp
  * c. estuary
    d. bog
    e. lagoon

34. The most sensitive organisms to pollution in the estuarine zone are:
    a. rodents
  * b. clams
    c. fish
    d. shrimp
    e. squid

35. Over ____% of the estuarine zone in the U.S. has been destroyed.
    a. 10
    b. 20
    c. 30
  * d. 40
    e. 50

36. This law allows for the establishment of national marine sanctuaries:
  * a. Coastal Zone Management Act
    b. National Wildlife Refuge System
    c. Wilderness Act
    d. Clean Water Act
    e. Coastal Barrier Resources Act

37. This law prohibits the expenditure of federal money for highway construction and other development on barrier islands:
    a. Coastal Zone Management Act
    b. National Wildlife Refuge System
    c. Wilderness Act
    d. Clean Water Act
  * e. Coastal Barrier Resources Act

38. This law provides protection of rivers from development:
    a. Wilderness Act
  * b. Wild and Scenic Rivers Act
    c. Clean Water Act
    d. National Environmental Policy Act
    e. Coastal Barrier Resources Act

39. These rivers are free of dams, largely undeveloped , accessible by roads, and of great scenic value:
    a. wild rivers.
  * b. scenic rivers
    c. recreational rivers
    d. free-flowing rivers
    e. national rivers

40. These structures are used by communities to reduce beach erosion by longshore currents:
    a. levees
    b. piers
  * c. jetties
    d. barrier islands
    e. reefs

132

15

Nonrenewable Energy Sources

Chapter Outline

Energy Use: Our Growing Dependence on Nonrenewable Fuels
Global Energy Consumption

Fossil Fuels: Analyzing Our Options
Crude Oil
Natural Gas
Coal
Impacts of Coal Production: Surface Mining
Impacts of Coal Production: Underground Mines
Impacts of Coal Combustion
Oil Shale
Tar Sands
Coal Gasification and Liquefaction

Fossil Fuel: Meeting Future Demand
Oil: The End is Near
Natural Gas: A Better Outlook
Coal: The Brightest Outlook, the Dirtiest Fuel

Nuclear Energy
Atoms: The Building Blocks of Matter
Radioactive Atoms
Radiation
Understanding Nuclear Fission
Fission By-products
Nuclear Power: The Benefits/The Drawbacks
Waste Disposal
Breeder Reactors
Nuclear Fusion

General Guidelines for Creating a Sustainable Energy System

Establishing Priorities
Short-term goals
Long-term Goals

Key Terms

energy system	crude oil	natural gas
coal	surface mining	contour strip mines
area strip mines	black lung disease	acid mine drainage
fly ash	smokestack scrubber	bottom ash
oil shale	kerogen	shale oil
net energy analysis	net energy yield	surface retort
in situ retorting	tar sands	bitumen
coal gasification	synfuels	coal liquefaction
ultimate production	nuclear energy	atom
nucleus	elements	isotopes
radiation	radionuclides	alpha particle
beta particle	gamma rays	X-rays
nuclear fission	nuclear fusion	daughter nuclei
light water reactors	fuel rods	reactor core
breeder reactor	magnetic confinement	energy quality

Objectives

1. List the major fossil fuels and discuss their environmental impacts.
2. Discuss the availability of fossil fuels in the future.
3. Define the following terms: "atom", "nucleus", "elements", and "isotopes".
4. Define the following terms: "nuclear fission", "nuclear fusion", "fuel rods", "reactor core" and "daughter nuclei".
5. List the different kinds of ionizing radiation.
6. Discuss the advantages and disadvantages of nuclear power.
7. Discuss the feasibility of using breeder reactors in place of conventional nuclear reactors.
8. List some guidelines for creating a sustainable energy system.
9. Summarize the energy quality of available fuels.

Lecture Outline

I. Energy Use: Our Growing Dependence on Nonrenewable Fuels
   A. U.S. Energy Consumption
      1. In the past century, wood was our major fuel; today oil, natural gas, and coal provide most of our energy.
   B. Global Energy Consumption
      1. Global Energy Consumption - Renewable fuels provide a significant portion of the energy sources in developing nations, whereas industrialized nations rely principally on nonrenewable energy sources.
      2. The U.S. accounts for 5.6% of the world's population but uses 25% of its energy; sheer wastefulness and inefficiency are partially to blame for this discrepancy.

II. Fossil Fuel: Analyzing Our Options - The energy system impacts human and ecosystem health at several points: exploration extraction, processing, distribution, and end use.
   A. Crude Oil
      1. Crude oil provides many useful fuel and nonfuel byproducts.
      2. Oil is extracted from deep wells on the seafloor and land. Often, natural gas is found along with oil
      3. The major impacts on the environment come from oil spills and from combustion of oil and its derivatives.
   B. Natural Gas - While not an ideal fuel, natural gas has few environmental impacts compared to other fossil fuels.
   C. Coal
      1. Strip mining and tunnel mining for coal have had considerable environmental impacts on large regions of the U.S.; impacts include erosion, habitat loss, siltation, acid drainage, and land subsidence.
      2. Transporting and burning coal produces a variety of air and water pollution and solid waste pollution problems.
   D. Oil Shale
      1. Oil Shale - Oil shale contains kerogen which can be heated to shale oil and refined into petroleum-like products.
      2. Though abundant and versatile, shale oil is expensive, inefficient, and causes extreme environmental damage from mining and refining.
   E. Tar Sands
      1. Bitumen in tar sands can be refined into petroleum substitutes; however, it is economically and environmentally costly, and supplies are limited.
   F. Coal Gasification and Liquefaction
      1. Costs air and water pollution, and net energy inefficiency make coal gasification and liquefaction low priority energy forms

III. Fossil Fuel: Meeting Future Demands
   A. Oil: The End Is Near - It appears that oil supplies may only last 20 to 40 years.
   B. Natural Gas: A Better Outlook - The outlook for natural gas better than coal with global reserves lasting about 60 years.
   C. Coal - Coal is the world's most abundant fossil fuel. The recoverable reserves would last 1700 years at the current rate of consumption. World proven reserves are estimated to last 200 years at the current rate of consumption.

IV. Nuclear Energy - Nuclear energy provides a growing percentage of the world's energy demands. Nuclear Fission produces heat by splitting fissile atoms such as U-235.
   A. Atoms: The Building Blocks of Matter
      1. The atom is composed of two parts: the *nucleus* and the *electron cloud*.
   B. Radioactive Atoms
      1. Elements - Elements are the purest form of matter. Not all atoms of every element are identical. All atoms of the same element contain the same number of *protons*, the positively charged particles in the nucleus. Some atoms of the same element have different numbers of neutrons; these atoms are called *isotopes*.
      2. Excess neutrons in some isotopes make them unstable and they emit radiation and are called *radioactive elements*.
   C. Radiation
      1. Radionuclides naturally emit radiation in one of three forms: alpha particles, beta particles, or gamma rays; X rays are not naturally occurring.

135

2. Ionizing radiation in any of the above forms damages living tissues by creating charged atoms, ions, which are highly reactive.
D. Understanding Nuclear Fission
   1. Fission - Fission is the splitting of atoms.
   2. Nuclear Fission - Inside a nuclear reactor U-235 atoms undergo fission when they are struck by neutrons. When the atoms split after being struck by the neutrons they release large amounts of energy.
E. Fission By-products - Uranium atoms that undergo fission release additional neutrons causing additional fission and heat. The chain reaction that is started is regulated by the use of control rods.
F. Nuclear Power: The Benefits
   1. Nuclear power does not contribute to the greenhouse effect or acid rain.
   2. It requires less surface mining and has lower fuel, transportation costs and impacts than does coal.
G. Nuclear Power: The Drawbacks
   1. Health Effects of Radiation - The effects of radiation depend to a large extent on the amount, exposure time, and type of radiation. Within a population, individuals in certain developmental stages are most susceptible to damage; similarly, certain cell types within an individual are more vulnerable dam others to ionizing radiation
   2. Health Impacts of High-Level Radiation - Though variable due to a number of factors, health effects of radiation include death, radiation sickness, and a host of delayed effects, including cancer and sterility.
   3. Health Impacts of Low-Level Radiation. Low-level doses are thought to increase cancer and leukemia rates, genetic defects, increase aging, and perhaps suppress natural immunity.
   4. Bioaccumulation - Bioaccumulation refers to the accumulation of radionuclides within tissues and the increasing concentration of those materials at higher trophic levels.
   5. Reactor Safety
      a. Attempts, such as the Rasmussen Report, to statistically assess nuclear power risks are fraught with difficulties and largely discredited today.
      b. Variables, including human error, terrorism, and sabotage make nuclear reactor safety assessment difficult.
      c. Accidents such as those at Three Mole Island and Chernobyl feed skepticism about industry claims of reactor safety.
   6. Waste Disposal
      a. Radioactive wastes are generated at several points in the fuel cycle; these wastes must be isolated for thousands of years.
      b. Disposal techniques and repositories for high-level radioactive wastes have not yet been developed despite the continued production of such wastes.
   7. Social Acceptability and Cost
      a. When all planning, construction, licensing, operating, maintenance, and decommissioning costs are factored in, nuclear power is our most expensive energy source today.
      b. High costs, safety concerns, and mismanagement have combined to lower the social acceptability of nuclear power relative to other energy sources.
   8. Proliferation of Nuclear Weapons
      a. The Link between nuclear power and nuclear bombs is well established.
H. Breeder Reactors

        1. Breeders convert an abundant material (U-238) into plutonium, which can then be extracted to fuel other reactors; thus, breeders can theoretically multiply our nuclear fuel supply.

        2. Drawbacks to breeder programs include cost, safety concerns, and plutonium's toxicity and desirability for nuclear weapons and terrorist explosives.

  I. Nuclear Fusion

        1. Nuclear fusion involves combining nuclei of light elements.

        2. Numerous technical problems related to the operating design and reactants products will delay commercial feasibility by several decades.

V. General Guidelines for Creating a Sustainable Energy System

  A. Sustainable energy future would incorporate the following: positive net energy production, energy matched in quality to the task, maximized efficiency and safety, minimized pollution, abundant and renewable energy sources, and affordability.

VI. Establishing Priorities

  A. Short-Term Goals (within 20 years)

        1. Efficiency and renewability must be the major goals of near-term energy planning.

  B. Long-Term Goals (within 50 years)

        1. A renewable replacement for natural gas must be found soon.

        2. A new policy for a responsible, sustainable energy program will almost certainly necessitate abandonment of old energy allegiances and customs.

Suggestions for Presenting the Chapter

- A visit to a local power generation facility is an excellent activity for this chapter
- Instructors should stress that our current dependence on fossil fuels is not sustainable activity. Currently, our energy supply system is unsustainable and we must move to sustainable energy sources.
- A discussion of the pros/cons of nuclear energy is an interesting class activity. This is particularly true if you have a nuclear power plant nearby.
- Instructors should emphasize why we need to set new goals for creating a sustainable energy system.
- Recommended web sites:

The Office of Surface Mining/U.S. Dept. of Interior: http://www.osmre.gov/
Bureau of Reclamation/U.S. Dept. of Interior: http://www.usbr.gov/main/index.html
U.S. Dept. of Energy/Sources and Production: http://www.energy.gov/sources/index.html
Exxon Valdez Oil Spill Restoration: http://www.oilspill.state.ak.us/
Nuclear Regulatory Commission: http://www.nrc.gov/

True/False Questions

1. ___Crude oil or petroleum is a thick liquid containing many combustible hydrocarbons.
2. ___40% of the energy used in the United States comes from oil.
3. ___Natural gas is a fossil fuel that does not burn as cleanly as coal.
4. ___Natural gas is a mixture of low-molecular weight hydrocarbons, mostly methane.
5. ___Coal is the most abundant fossil fuel.
6. ___Coal is a liquid fuel that is heavier in density than natural gas.
7. ___The Surface Mining Control and Reclamation Act was enacted in 1956.
8. ___Strip mines have very minimal impacts on the local environment.
9. ___Fly ash is mineral matter released during the operation of wind generators.
10. ___Sulfur dioxide is one pollutant produced during the combustion of coal.
11. ___Many abandoned coal mines in the East leak sulfuric acid into streams.
12. ___Acid mine drainage increases the fertility of rivers it leaks into.
13. ___Sulfur dioxides in stack gases can be removed using a scrubber.
14. ___Oil shale is an igneous rock containing a combustible organic material.
15. ___Tar sands contain a combustible organic material known as synfuel.
16. ___Gamma rays are a high-energy form of radiation with no mass and no charge.
17. ___Nuclear reactors that are cooled with water are called light water reactors.
18. ___Fission reactions in the nuclear reactor are held in check by using a breeder reactor.
19. ___The sun generates energy by nuclear fusion.
20. ___Nuclear power has become a socially unacceptable form of electricity in part because of the high costs associated with all phases of operation.

True/False Key:

1. T 2. T 3. F 4. T 5. T 6. F 7. F 8. F 9. F 10. T 11. T 12. F 13. T 14. F 15. F 16. T 17. T 18. F 19. T 20. T

Fill-in-the-Blank Questions

1. Breeder reactors are designed to produce energy and radioactive fuel from uranium - ___.
2. Nuclear _____ occurs when two hydrogen nuclei join.
3. The half-life of plutonium-239, a waste product of nuclear reactors, is _____ years.
4. The DOE site selected for the long-term storage of nuclear wastes is _____ Mountain in Nevada.
5. Radionuclides often build-up in tissues, a process called _____.
6. Fission reactions in nuclear reactors are held in check by the _____ rods.
7. The fuel used in the typical fission nuclear reactor is _____.
8. _____ is the splitting of atoms.

9. _____ particles consist of two protons and two neutrons.
10. Unstable, radioactive nuclei are called _____.
11. The nucleus of an atom contains _____ and neutrons.
12. The negatively charged _____ orbit around the nucleus of an atom.
13. Coal _____ produces combustible gases that can supplement natural gas.
14. Tar sands contain a combustible organic material known as _____.
15. Oil _____ is a sedimentary rock containing a combustible material known as kerogen.
16. Mined shale is crushed and heated in a large vessel called a surface _____.
17. Collapsing mines also cause _____, a sinking of the surface over the mine.
18. Underground mines produce enormous quantities of wastes called _____.
19. Fly ash is a mineral material that makes up 10-30% of the weight of uncleaned _____.
20. Natural gas is a mixture of low-molecular-weight _____, mostly methane.

Fill-in Key:

1. 238
2. fusion
3. 24,000
4. Yucca
5. bioaccumulation
6. control
7. uranium-235
8. fission
9. alpha
10. radionuclides
11. protons
12. electrons
13. gasification
14. bitumen
15. shale
16. retort
17. subsidence
18. tailings
19. coal
20. hydrocarbons

Multiple Choice Questions

1. This fuel accounted for 40% of the energy consumed in the U.S. in 1995:
   a. natural gas
   b. ethanol
   c. coal
   * d. oil
   e. wood

2. Renewable energy sources supplied about ___% of our energy needs in 1995.
   a. 3
   * b. 6
   c. 12
   d. 32
   e. 48

3. Over half of the energy in the U.S. Is consumed by:
   a. transportation
   b. residential use
   * c. business and industry
   d. government
   e. recreational use

4. Worldwide the biggest users of energy are Americans, who make up about 5.6% of the world's population but consume about ___% of its energy.
   a. 10
   * b. 25
   c. 50
   d. 75
   e. 90

5. This organic mixture is a thick liquid and is a source for many fuel and nonfuel products:
   a. natural gas
   b. methane
   * c. crude oil
   d. kerosene
   e. tar

6. This fuel is a mixture of low-molecular weight hydrocarbons, mostly methane:
   a. tar
   b. ethanol
   c. kerosene
   d. fuel oil
   * e. natural gas

7. This is the most abundant fossil fuel but is very environmentally costly:
   a. tar
   b. oil
   * c. coal
   d. natural gas
   e. ethanol

8. Which of the following is an environmental impact of coal extraction?
   a. Destruction of wildlife habitat.
   b. Soil erosion from mine sites.
   c. Aquifer depletion and pollution.
   d. Acid mine drainage.
   * e. all of the above

9. Which of the following is an environmental impact of oil mining?
   a. Offshore leaks and blowouts causing water pollution.
   b. Soil erosion.
   c. Acid mine drainage.
   d. Black lung disease
   e. Underground mine accidents.

10. In mining, the overburden is the:
    a. overhead material in an underground mine.
    b. the unusable rock in an ore deposit.
    * c. rock and dirt that must be removed in surface mining.
    d. tailings left after ore extraction.
    e. surface erosion that occurs after surface mining.

11. In flat areas, surface mining of coal is done using:
    *   a. area strip mines
        b. contour strip mines
        c. underground mines
        d. dredges
        e. drilling

12. Area strip mines create which of the following environmental impacts?
        a. Soil erosion.
        b. Destruction of wildlife habitat.
        c. Destroy grazing land.
        d. Create unpleasant areas.
    *   e. all of the above

13. Coal in mountainous terrain is extracted using:
        a. strip mines
        b. surface mines
    *   c. underground mines
        d. drilling
        e. dredges

14. Underground mining can cause the collapse of the overlying ground, a process called:
        a. sublimation
        b. block faulting
    *   c. subsidence
        d. draglining
        e. infiltration

15. Acid mine drainage:
        a. makes waters extremely acidic.
        b. inhibits bacterial decay of matter in water.
        c. kills plants and animals in the water.
        d. leaches toxic metals into the water.
    *   e. all of the above

16. Most of the streams polluted by acid mine drainage in the U.S. are located in:
        a. Alaska
        b. Colorado
        c. Wyoming
    *   d. Appalachia
        e. California

17. About _____ of the coal in the U.S. is used to generate electricity.
        a. one-eighth
        b. one-fourth
        c. one-third
        d. one-half
    *   e. two-thirds

18. One of the major air pollutants of coal combustion is:
    *   a. sulfur dioxide
        b. water
        c. methane
        d. ozone
        e. bromine

19. This sedimentary rock contains the organic material kerogen:
        a. coal
        b. sandstone
        c. tar sand
    *   d. oil shale
        e. limestone

20. Sand deposits containing large amounts of combustible material known as bitumen are called:
        a. sandstone
        b. oil shale
    *   c. tar sand
        d. limestone
        e. shale oil

21. The production of synthetic oil from coal is called:
        a. coal gasification
    *   b. coal liquefaction
        c. cracking
        d. synthesis
        e. refining

22. Current oil supplies are predicted to last only _____ years.
        a. 5 to 10
        b. 10 to 20
    *   c. 20 to 40
        d. 40 to 80
        e. 100 to 200

23. The U.S. has about _____% of the world's proven coal reserves.
   a. 10
   b. 20
* c. 30
   d. 40
   e. 50

24. Air pollution control technologies for coal burning do little to prevent global climate change because they do not reduce emission of this substance:
   a. nitrous oxide
   b. sulfur dioxide
   c. water
* d. carbon dioxide
   e. ozone

25. Most nuclear reactors today are powered by:
* a. U-235
   b. U-238
   c. Pu-239
   d. Helium
   e. heavy water

26. The nucleus of an atom contains:
   a. electrons only
   b. neutrons only
   c. electrons and neutrons
* d. protons and neutrons
   e. electrons and neutrons

27. Most of the mass of the atom (99.9%) is found in the:
   a. electron
   b. neutron
   c. proton
* d. nucleus
   e. electron cloud

28. This form of radiation contains two protons and two neutrons:
* a. alpha particle
   b. beta particle
   c. gamma ray
   d. cosmic rays
   e. ultraviolet radiation

29. This form of radiation is equivalent to an electron except they are more energetic:
   a. alpha particle
* b. beta particle
   c. gamma rays
   d. cosmic rays
   e. X-rays

30. All commercial nuclear reactors use:
   a. nuclear fusion
* b. nuclear fission
   c. breeder reactors
   d. supercooling
   e. magnetic confinement

31. One environmental advantage of nuclear power is that is does not produce _____ pollution.
   a. water
   b. noise
* c. air
   d. chemical
   e. radiation

32. Which of the following responses is a disadvantage of nuclear power:
   a. nuclear waste disposal
   b. thermal pollution of water
   c. health impacts of radiation
   d. limited supplies of uranium
* e. all of the above

33. Breeder reactors could extend the supply of fissionable nuclear material considerably since they can use:
   a. U-235
* b. U-238
   c. Strontium-90
   d. magnesium
   e. boron

34. _____ occurs when two hydrogen nuclei join to form a helium nucleus.
   a. Nuclear fission
* b. Nuclear fusion
   c. Magnetic confinement
   d. Containment
   e. Kinetic energy release

35. Which of the following forms of energy provides the highest *energy quality?*
    a. geothermal
    b. biomass
    c. natural gas
  * d. electricity
    e. manure

36. Nuclear chain reactions are driven by the release of _____ from the split nuclei of radioactive atoms.
    a. electrons
  * b. neutrons
    c. protons
    d. cosmic rays
    e. X-rays

37. What state is the Yucca Mountain site located (this is the proposed site for a nuclear waste repository):
    a. California
  * b. Nevada
    c. Arizona
    d. Utah
    e. Idaho

38. Nuclear fusion is being pursued as a possible source of energy for the future because its fuel supply can be extracted from:
    a. plutonium
    b. uranium
    c. sodium
    d. boron
  * e. seawater

39. The world's supply of uranium-235 used in light water reactors will last about _____ years.
  * a. 100
    b. 200
    c. 300
    d. 400
    e. 500

40. This source of radiation does not originate from naturally occurring unstable nuclei:
    a. alpha particles
    b. beta particles
    c. gamma rays
  * d. X-rays
    e. cosmic rays

16

Foundations of a Sustainable Energy System: Conservation and Renewable Energy

Chapter Outline

Energy Conservation: Foundation of a Sustainable Energy System
Economic and Environmental Benefits of Energy Conservation
Energy-Efficiency Options
The Potential of Energy Efficiency
Promoting Energy Efficiency
Roadblocks to Energy Conservation

Renewable Energy Sources
Solar Energy Options
Wind Energy
Biomass
Hydroelectric Power
Geothermal Energy
Hydrogen Fuel

Renewable Energy: What's Its Potential?
Assessing the Supplies
Is a Renewable Energy Supply System Possible?
Economic and Employment Potential of the Sustainable Energy Strategy

Key Terms

energy conservation
cogeneration
solar energy
earth-sheltered home
photovoltaics
wind energy
hydroelectric power
hydrothermal convection zones
total resources

energy efficiency
least-cost planning
passive solar heating
solar collectors
solar thermal electric
biomass
magma
geopressurized zones
accessible resources

gross national product
renewable energy
active solar systems
flat plate collectors
copper cricket
geothermal energy
hydrogen fuel
hot-rock zones

Objectives

1. Discuss the economic and environmental benefits of energy conservation.
2. List some ways we can take advantage of energy efficient technologies.
3. Summarize how we can personally reduce our energy use.
4. Discuss how energy efficiency can be promoted.
5. List the types of renewable energy currently available.
6. Discuss the advantages and disadvantages of each type of renewable energy.
7. Discuss the feasibility of renewable energy replacing our dependence on fossil fuels.
8. Define the terms: "total resources" and "accessible resources".

Lecture Outline

I. Energy Conservation: Foundation of a Sustainable Energy System - Energy waste is economically, environmentally, and socially irresponsible.
  A. Economic and Environmental Benefits of Energy Conservation
    1. Conservation requires us to avoid unnecessary and inefficient use of energy.
    2. Energy efficiency contributes to a strong economy and minimizes fuel system-related pollution and resource depletion.
  B. Energy-Efficiency Options
    1. Conservation techniques include increasing efficiency of fuel use, better technology in buildings, appliances, and industrial processes, and lifestyle changes which reduce personal energy consumption.
  C. The Potential of Energy Efficiency
    1. Transportation Savings
      a. Improvements in automobile fuel efficiency hold great potential for reducing energy consumption and pollution, especially in the U.S.
    2. Buildings
      a. A reversal of current incentives which tend to impede the implementation of energy efficient building practices could substantially reduce energy consumption in the U.S.
    3. Industries
      a. Recycling and increasingly efficient use of energy for industrial processes will boost profits and reduce environmental impacts of production in the U.S.
  D. Promoting Energy Efficiency
    1. Incentives for implementing existing conservation techniques and developing more include taxes, government mandated efficiency standards, pricing, and least-cost planning.
  E. Roadblocks to Energy Conservation
    1. Government funding bias and consumption incentives, initial investment cost, misinformation regarding fossil fuel supplies, and successful lobbying by the energy industry have combined to thwart the timely transition to a sustainable energy system in the U.S.

II. Renewable Energy Sources
  A. Solar Energy Options

1. Passive Solar Heating
   a. This approach is to design buildings which passively utilize the sun's energy for space heating and lighting.
2. Active Solar
   a. These systems effectively and efficiently heat and cool air and/or water for residential and commercial buildings.
3. Photovoltaics
   a. Solar cells can generate electricity from sunlight; these are especially useful today in situations where other fuels are expensive or impractical.
4. Solar Thermal Electric
   a. The sun's energy can be captured and concentrated in a variety of ways to generate electricity directly.
5. Pros and Cons of Solar Energy
   a. Solar energy is our most flexible and widely available, low-impact and nondepletable energy source.
   b. Free fuel and low maintenance costs make solar economically appealing to forward-looking homeowners.
   c. Limitations include solar's intermittency and resulting need for storage or back-up systems.

B. Wind Energy - The potential for wind-generated electricity, heat, and other direct applications is enormous.
   1. Pros and Cons of Wind Energy
      a. Wind energy has most of the same advantages and disadvantages as solar energy, with the added disadvantage of greater visual and audio impact. Some wind generators may interfere with television reception and microwave communications.

C. Biomass - Energy from organic matter of plants, biomass, is significant in many parts of the world today, and can become more so in the U.S. if given a push by the government.
   1. Pros and Cons of Biomass
      a. While abundant and widely available, drawbacks include increased competition between people for food and energy supplies if more resources are devoted to fuel rather than food farms.

D. Hydroelectric Power
   1. Pros and Cons of Hydroelectric Power
      a. This energy source is renewable, relatively nonpolluting, and inexpensive, yet creation and maintenance of large dams and reservoirs have severe localized ecological impacts.

E. Geothermal Energy
   1. Pros and Cons of Geothermal Energy
      a. Hydrothermal convection, geopressure, and hot-rock zones are geothermal resources which are renewable and fairly clean but which have limited potential for future use.

F. Hydrogen Fuel
   1. Pros and Cons of Hydrogen Fuel
      a. Made from splitting water molecules, hydrogen gas is renewable, low polluting, and versatile; however, its low net-energy yield must be improved in order to justify widespread development.

III. Renewable Energy: What's Its Potential
   A. Assessing the Supplies

1. Renewable energy makes up the bulk of total and accessible energy resources worldwide.
   B. Is a Renewable Energy Supply System Possible?
      1. The first step in the transition to energy sustainability is increased efficiency; renewable fuels, and increasing use of solar and wind energy and biomass. Fossils fuels will begin to be replaced as they become too environmentally and economically costly.
   C. Economics and Employment Potential of the Sustainable Energy Strategy
      1. Economics and convention, both heavily influenced by government subsidies, give nonrenewable fuels the advantage today.
      2. When all casts are internalized we find that conservation and renewable energy systems are economically preferable and have a net positive impact on employment.

Suggestions for Presenting the Chapter

- Instructors might arrange a class visit to a passive solar home. Many contractors are happy to show how the homes are built and what features are available for energy efficiency.
- Instructors should stress that the technologies for conversion to sustainable alternatives to our energy needs are already available and cost effective. Conservation and efficiency should be stressed as areas of immediate concern in our everyday lives.
- Recommended web sites:

  National Renewable Energy Laboratory/U.S. Dept. of Energy: http://www.nrel.gov/
  Rocky Mountain Institute: http://www.rmi.org/
  Energy Star Program: http://www.energystar.gov/
  International Rivers Network/Three Gorges Dam: http://irn.org/programs/threeg/
  Center for Renewable Energy & Sustainable Technology:
          http://solstice.crest.org/common/crestinfo.shtml

True/False Questions

1. ___ The National Appliance Conservation Act established energy-efficiency standards for all new appliances.
2. ___ Photovoltaics generate electricity by using hydrogen gas as a fuel.
3. ___ Solar thermal electric facilities heat water using sunlight.
4. ___ Active heating and cooling systems employ solar reflectors.
5. ___ Biomass is organic matter such as wood or crop wastes.
6. ___ Hydroelectric power is a nonrenewable resource that uses hydrogen gas.
7. ___ The Earth harbors an enormous amount of heat or geothermal energy.
8. ___ In Iceland, 65% of the homes are heated by the Earth's heat.
9. ___ Hydrogen is a nonrenewable fuel that is not a good choice to replace petroleum fuels.
10. ___ Geopressurized zones are aquifers that are trapped by impermeable rock strata and heated by underlying magma.

11. ___ Geothermal energy is heavily concentrated in a so called ring of fire encircling the Aral Sea.
12. ___ Wind energy is clean, abundant, and fairly inexpensive energy.
13. ___ Biomass supplies about 50% of the world's energy.
14. ___ Earth-sheltered houses are partly or entirely underground to take advantage of the insulative properties of the soil.
15. ___ Over one-fourth of the energy consumed in the United States is used in transportation.
16. ___ A "feebate" is a tax paid by those who buy gas-guzzling cars and a rebate given to those who purchase energy-efficient autos.
17. ___ Hot-rock zones are regions where bedrock is heated by magma.
18. ___ Ethanol and methane are two fuels that can be produced from biomass.
19. ___ Passive solar homes are kept warm by residual heat that radiates from heat-absorbent materials (thermal mass).
20. ___ Photovoltaic cells are made of silicon and other materials.

True/False Key:

1. T 2. F 3. T 4. F 5. T 6. F 7. T 8. T 8. T 9. F 10. T 11. F 12. T 13. F 14. T 15. T 16. T 17. T 18. T 19. T 20. T

Fill-in-the-Blank Questions

1. Planting _____ trees around homes can reduce summer cooling costs.
2. In industrial _____ waste heat from one process is captured and used in another.
3. Huge cuts in energy demand can be made by applying energy _____ measures.
4. About one-third of the energy consumed in the U.S. is consumed in _____.
5. The National Appliance Conservation Act was passed by Congress in _____.
6. _____ planning requires power companies to select the least costly way of providing electricity.
7. Well-designed passive systems can provide ___% of a home's space heating.
8. Solar energy is considered a _____ energy source.
9. The earth-_____ house is built partly or entirely underground to take advantage of the insulative properties of soil.
10. Active heating and cooling systems employ solar _____.
11. _____ provide a way of generating electricity from sunlight.
12. Solar _____ electric facilities heat water using sunlight.
13. _____ solar energy stores heat in thermal mass.
14. _____ are produced by solar energy and can be used to generate electricity.
15. _____ is organic matter such as wood or crop wastes that can be burned or converted into gaseous or liquid fuels.
16. Biomass supplies about ___% of the world's energy.
17. Unlike fossil fuels, biomass does not pollute the atmosphere with _____ dioxide.
18. _____ power is a renewable resource usually tapped by damming streams and rivers.

19. _____ convection zones are places where magma intrudes into the Earth's crust and heats rock that contains large amounts of groundwater.

20. _____ zones are aquifers that are trapped by impermeable rock strata and heated by underlying magma.

Fill-in Key:
1. shade
2. cogeneration
3. efficiency
4. buildings
5. 1987
6. Least-cost
7. 100
8. renewable
9. sheltered
10. collectors
11. Photovoltaics
12. thermal
13. Passive
14. winds
15. Biomass
16. 20
17. carbon
18. hydroelectric
19. Hydrothermal
20. Geopressurized

Multiple Choice Questions

1. When waste heat from one process is captured and used in another this is called:
   a. dovetailing
   b. codependency
   * c. cogeneration
   d. heat exchange
   e. counter current exchange

2. The World Resources Institute estimates that the world could meet 90 % of its new energy needs between 1987 and 2020 by:
   a. using solar power.
   b. increasing use of coal.
   c. using geothermal energy.
   d. building more nuclear power plants.
   * c. using energy more efficiently.

3. About ___% of all delivered energy is needed as heat.
   a. 18
   b. 28
   c. 38
   d. 48
   * e. 58

4. Solar energy from the sun is expected to last some _____ billion years.
   a. 1
   * b. 2
   c. 3
   d. 4
   e. 5

5. The average automobile in the U.S. gets _____ miles per gallon of gasoline.
   a. 18
   * b. 28
   c. 38
   d. 48
   e. 58

6. About one-third of the energy in the U.S. is consumed in:
   a. automobiles
   * b. buildings
   c. industry
   d. government
   e. transportation

7. Which of the following responses would promote energy efficiency?
   a. education
   b. taxes on fossil fuels
   c. feedback systems
   d. government-mandated efficiency programs
   * e. all of the above

8. Least-cost planning is used by many states to encourage efficiency in which of the following areas:
   a. transportation
   b. sewage disposal
   * c. power generation
   d. construction
   e. landfill management

9. Flat plate collectors are used in:
   a. passive solar homes
   b. earth-sheltered homes
   * c. active solar homes
   d. log homes
   e. concrete

10. This type of home uses sunlight to heat water and produce steam to generate electricity:
    a. passive solar
    b. earth-sheltered
    * c. solar thermal electric
    d. wind powered
    e. solar retrofit

11. The National Appliance Conservation Act:
    * a. Established energy efficiency standards for all new appliances.
    b. Regulates the disposal of ozone depleting chemicals from appliances.
    c. Orders states to dispose of appliances properly.
    d. Requires labeling of all appliances with warnings about environmental hazards.
    e. Specifies what must be done by companies to comply with ozone regulations.

12. Buildings designed to capture energy from the sun using interior walls and floors to provide space heat are using _____ heating.
    a. active solar
    * b. passive solar
    c. photovoltaic
    d. solar thermal electric
    e. solar retrofit

13. Electricity generated when sunlight strikes a solar panel is an example of:
    a. passive solar
    b. active solar
    * c. photovoltaics
    d. solar thermal electric
    e. solar retrofit

14. Earth-sheltered homes take advantage of the insulative properties of:
    a. concrete
    * b. soil
    c. tile
    d. water
    e. wood

15. Places where magma protrudes into the earth's crust and heats rock that contain large amounts of groundwater are called:
    * a. hydrothermal convection zones
    b. geopressurized zones
    c. hot-rock zones
    d. thermal vents
    e. hot springs

16. Homeowners can add solar greenhouses to existing conventional homes, this is called a _____ home.
    a. passive solar
    b. active solar .
    c. earth-sheltered
    * d. solar retrofit
    e. greenhouse

17. A primary element found in many photovoltaic panels is:
    a. water
    b. air
    c. carbon
    * d. silicon
    e. boron

18. About ___% of the sun's energy striking the Earth is converted to wind.
    * a. 2
    b. 10
    c. 20
    d. 40
    e. 60

19. Which of the following responses is **not** one of the advantages of wind energy? Wind energy is:
    a. cheap.
    b. clean.
    c. renewable.
    d. cost-competitive with other forms of energy.
    * e. dependent on the availability of the wind.

20. The use of wood or crop wastes as a source of energy is called:
    a. geothermal energy
    * b. biomass
    c. synfuel
    d. fossil fuels
    e. hydrogen fuel

21. The molten rock beneath the Earth's crust is:
    a. granite
    b. bedrock
    * c. magma
    d. plasma
    e. molten iron

22. Aquifers that are trapped by impermeable rock strata and heated by underlying magma are called:
    a. hydrothermal convection zones
    * b. geopressurized zones
    c. hot-rock zones
    d. thermal vents
    e. hot springs

23. Regions where bedrock is heated by magma are called:
    a. hydrothermal convection zones
    b. geopressurized zones
    * c. hot-rock zones
    d. thermal vents
    e. hot springs

24. Hydrogen gas has an advantage over fossil fuels since its oxidation does not release _____, the primary pollutant associated with global warming.
    a. ozone
    b. methane
    * c. carbon dioxide
    d. chlorofluorocarbons
    e. water

25. Which of the following is a personal action you can take to reduce energy use?
    a. Turn down the thermostat on the hot water heater.
    b. Insulate your walls and ceilings.
    c. Heat only used areas.
    d. Use energy-efficient appliances.
    * e. all of the above

26. If you study the land use of electricity generating technologies you find that _____ uses less land than any other energy source.
    a. coal (includes mining)
    b. solar thermal electric
    c. photovoltaics
    d. wind energy
    * e. geothermal energy

27. When federal subsidies paid by taxpayers are included, the cost of U.S. oil is estimated to be about _____ a barrel compared to the market price of about $22 per barrel.
    a. $25 to $50
    b. $50 to $75
  * c. $100 to $200
    d. $300 to $400
    e. $400 to $500

28. Active solar heating and cooling systems use _____ mounted on rooftops.
    a. wind machines
    b. radiators
    c. heat sinks
  * d. solar collectors
    e. parabolic mirrors

29. Solar thermal electric systems use _____ to direct sunlight onto an oil-filled tube.
    a. wind machines
    b. radiators
    c. heat sinks
    d. solar collectors
  * e. parabolic aluminum reflectors

30. When sunlight strikes a photovoltaic cell, _____ are ejected from silicon atoms and flow through electrical wires.
    a. protons
    b. neutrons
  * c. electrons
    d. photons
    e. gamma rays

31. Which of the following is a disadvantage of wind energy?
    a. Wind does not blow all the time..
    b. Individual windmills and wind farms are eyesores.
    c. Large wind generators may be noisy.
    d. Wind generators may impair television signals.
  * e. all of the above.

32. Biomass can help developed nations reduce their dependence on _____ energy resources.
    a. solar
    b. wind
    c. geothermal
  * d. nonrenewable
    e. hydroelectric

33. The world's leader in hydroelectric production is:
    a. France
    b. Brazil
  * c. United States
    d. China
    e. Italy

34. Which of the following would be good strategies for increasing hydropower today?
    a. Build more dams and facilities.
    b. Tear down old facilities and build more efficient ones.
    c. Increase the capacity of existing facilities by adding more turbines.
    d. Install turbines on dams built for flood control, recreation, and water supplies.
  * e. c and d above

35. Which of the following is a renewable resource?
    a. solar energy
    b. wind energy
    c. hydroelectric power
    d. geothermal energy
  * e. all of the above

36. Renewable energy resources account for _____% of the world's fuel supply.
    a. 15
    b. 30
    c. 45
    d. 60
  * e. 90

152

37. Renewable energy technologies employ more people than conventional technologies because they are:
   a. more costly than conventional technologies.
   b. more capital intensive.
* c. more labor-intensive.
   d. using cruder technology.
   e. less reliable and require more maintenance.

38. Electricity from coal-fired power plants costs _____ cents per kilowatt-hour.
   a. 1 to 2
* b. 5 to 7
   c. 8 to 12
   d. 15 to 20
   e. 25 to 30

39. Electricity from nuclear power plants costs _____ cents per kilowatt-hour.
   a. 1 to 2
   b. 5 to 7
* c. 8 to 12
   d. 15 to 20
   e. 25 to 30

40. Every $1 you invest in energy-efficiency measures will save you _____ in oil costs.
   a. $2
   b. $4
   c. $6
   d. $8
* e. $10

153

17

The Earth and Its Mineral Resources

Chapter Outline

Key Terms

minerals	ores	nonfuel minerals
metal-yielding minerals	industrial minerals	construction minerals
igneous rocks	sedimentary rocks	metamorphic rocks
infrastructure	strategic minerals	smelting
open pit mines	hydraulic mining	heap leaching
recycling	residence time	depletion allowance
reserves	manganese nodules	

Objectives

1. Classify the earth's nonfuel minerals.
2. List the three major classes or rocks.
3. Discuss which countries use most of the mineral resources globally.
4. Define the terms: "strategic minerals", "smelting", "open pit mines", "hydraulic mining", and "heap leaching".
5. Discuss the environmental problems caused by mining for minerals.
6. Summarize the steps necessary to create a sustainable system of mineral production.
7. Define the following terms: "recycling", "residence time", and "reserves".
8. Discuss the environmental concerns associated with the General Mining Act of 1872.
9. Discuss what personal actions can be taken to preserve the earth's minerals.

Lecture Outline

I. The Earth's Mineral Riches - The earth started as a molten mass, but the surface cooled, forming the crust.
   A. Minerals and Ores
      1. Geological processes often concentrate minerals in igneous rocks.
      2. An ore is a mineral deposit which can be economically mined or refined.
   B. Mineral Resources and Society
      1. Minerals are extremely important in our lives. Metals derived from ores are used in many products. Industrial minerals are used in fertilizers and concrete.
      2. More than 100 nonfuel minerals are traded in the world market.
      3. Among the most important are iron, aluminum, and copper.
   C. Who Consumes the World's Minerals?
      1. The developed nations consume a disproportionately large amount of the minerals marketed today.
   D. Import Reliance
      1. Large developed nations heavily depend on mineral imports; this dependence creates vulnerability because of political and economic instability of the exporting nations.
   E. Will There Be Enough?
      1. Concerns about long-term supplies of important minerals and rates of use are legitimate.
      2. Approximately 18 minerals will be 80% depleted by or before 2040.
   F. Environmental Impacts of Mineral Exploitation: A Brief Overview
      1. Ore extraction by open pit or hydraulic mining has severe environmental consequences including ecosystem destruction, aquatic deterioration from acid drainage and sedimentation, air pollution.
      2. Smelting, heap leaching, and other ore processing operations can have devastating environmental effects over vast areas.

II. Supplying Mineral Needs Sustainably
   A. Creating a Sustainable System of Mineral Production
      1. Recycling
         a. Recycling can dramatically slow our consumption of virgin ores by increasing a mineral's residence time.

b. Since recycling can never reach 100% for any mineral it cannot be the
　　　　solution to mineral depletion.
　2. Conservation
　　　a. Conservation can extend the lifetimes of many valuable mineral supplies.
　3. Promoting Conservation and Recycling
　　　a. Changes in legislation and tax incentives are needed to promote better use of
　　　　world mineral resources.
　4. Restoration and Environmental Protection
　　　a. Efforts must be made not only to protect the environment from current
　　　　mining impacts, but also to restore those areas already heavily impacted.

III. Expanding Reserves - Future demand cannot be completely satisfied by recycling and
　　conservation efforts. Some new minerals will need to be mined.
　A. Rising Prices, Rising Supplies
　　1. Short-term expansion is feasible, whereas long-term expansion is problematic.
　　2. The laws of supply and demand indicate that we can expand supplies by
　　　recovering less accessible reserves as price increases make this economically
　　　feasible.
　　3. In the long run, though, increased demand and diminishing supply must
　　　ultimately deplete a nonrenewable mineral resource.
　B. Technological Advances Expand Reserves
　　1. While helpful in increasing extraction efficiency, technological advances cannot
　　　make mineral resources infinite in supply.
　C. Factors That Reduce Supplies
　　1. Many factors, including energy and environmental constraints, can inhibit
　　　production and expansion of mineral reserves.
　D. Minerals From Outer Space and the Sea
　　1. Untapped mineral deposits await us in outer space, Antarctica, and the oceans --
　　　but only with enormous environmental damage, cost, effort, and risk.

IV. Finding Substitutes
　A. While substitutes undoubtedly can be found for some depleted minerals, they may
　　themselves be limited; and, in any case, recycling and conserving are preferable to
　　reliance on technological fixes.

V. Personal Actions
　A. Careful and reduced buying, adopting a low-impact lifestyle, and supporting
　　legislation aimed at increasing conservation and recycling will help initiate society-
　　wide changes necessary to slow mineral depletion

Suggestions for Presenting the Chapter

- A fieldtrip to a local mining sight, if available, is a good way to explore the concepts in this
  chapter.
- Instructors should stress that conservation and recycling are effective personal, societal and
  global sustainable activities that conserve the earth and its minerals.
- Instructors should emphasize the individual responsibility and personal impact involved with
  conservation and recycling. The importance of careful and reduced buying and the adoption
  of a low-impact lifestyle are pertinent issues associated with this chapter.

- Recommended web sites:

  U.S. Geological Survey: http://www.usgs.gov/
  Asbestos and Mining/Libby Montana Story: http://www.scn.org/~bh162/asbestos_libby.html
  Office of Surface Mining/U.S. Dept. of Interior: http://www.osmre.gov/
  Basics: Rocks & Minerals:
         http://geology.about.com/science/geology/library/bl/blbasics_roxmin.htm
  Fundamentals of Physical Geography:
         http://www.geog.ouc.bc.ca/physgeog/studyguide/studyguide.html

True/False Questions

1. ___ Minerals are largely inorganic substances such as sand, gravel and iron ore.
2. ___ Oil and coal are referred to as nonfuel minerals.
3. ___ An example of an industrial mineral would be lime.
4. ___ An example of a construction mineral would be copper ore.
5. ___ Basalt and granite are igneous rocks.
6. ___ Shale and sandstone are metamorphic rocks.
7. ___ The most abundant element's in the Earth's crust are oxygen, silicon, aluminum and magnesium.
8. ___ Most metal-yielding minerals come from sedimentary rocks.
9. ___ More than 100 nonfuel minerals are traded in the world market.
10. ___ The developed countries have 25% of the world's population and consume 75 % of its mineral resources.
11. ___ Melting ore to extract metal is called smelting.
12. ___ Hydraulic mining uses large hydraulic jacks to crush the raw ore.
13. ___ Heap leaching is used to extract gold from crushed ore.
14. ___ The time that a mineral remains in use is known as the residence time.
15. ___ The General Mining Act of 1872 permits miners to purchase federally owned land for very cheap prices.
16. ___ The depletion allowance provides a tax exemption to mining companies to compensate for declining ore reserves.
17. ___ The Resource Conservation and Recovery Act requires that mining smelters report their emissions to the EPA.
18. ___ Approximately 90% of American cars are recycled.
19. ___ Geologists use the term "infrastructure" to indicate deposits of minerals that they are fairly certain exist and that are feasible to mine at current prices.
20. ___ The United States stockpiles a three-year supply of strategic minerals such as bauxite, cadmium and graphite.

True/False Key:

1. T 2. F. 3. T 4. F 5. T 6. F 7. T 8. F 9. T 10. T 11. T 12. F 13. T 14. T 15. T 16. T 17. F 18. T 19. T 20. T

Fill-in-the-Blank Questions

1. Gold is sometimes extracted from crushed ore by spraying a cyanide solution of the piles, a process called _____ leaching.
2. Hillsides can be blasted with high-pressure water to extract ores from the soil by a process called _____ mining.
3. The United States stockpiles a _____-year supply of strategic minerals.
4. A concentrated deposit of minerals that can be mined and refined economically is called an _____.
5. Bridges, highways and buildings form the _____ of a nation.
6. The United States has about 6% of the world's population and consumes about ____% of its minerals.
7. Copper is used to make electrical wires and water _____.
8. Aluminum and copper ore are _____-yielding minerals.
9. Basalt and granite are _____ rocks.
10. _____ are largely inorganic substances such as sand, gravel and iron ore.
11. Coal and oil are _____ minerals.
12. Lime and pumice are _____ minerals.
13. Sand and gravel are _____ materials.
14. Solid aggregates that usually contain two or more different types of mineral are called _____.
15. The Earth started as a molten mass but the surface cooled, forming the _____.
16. Stone, clays and lime are _____ minerals.
17. Aluminum, zinc and silver are _____.
18. The process of melting ore to extract a metal is called _____.
19. The General Mining Act of _____ permits miners to purchase federally owned lands.
20. The Resource Conservation and Recovery Act places tight controls on _____ wastes.

Fill-in Key:

1. heap
2. hydraulic
3. three
4. ore
5. infrastructure
6. 20
7. pipe
8. metal
9. igneous
10. minerals
11. fuel
12. industrial
13. construction
14. rocks
15. crust

16. nonmetal
17. metals
18. smelting
19. 1872
20. hazardous

Multiple Choice Questions

1. Geologic processes often concentrate minerals in _____ rocks.
   a. sedimentary
   b. metamorphic
   * c. igneous
   d. bedrock
   e. surface

2. Basalt and granite are:
   a. sedimentary rocks
   b. metamorphic rocks.
   * c. igneous rocks
   d. metal-yielding minerals
   e. industrial minerals

3. Oil and coal are _____ minerals.
   a. metal-yielding
   b. industrial
   * c. fuel
   d. construction
   e. not

4. Gravel and sand are _____ minerals.
   a. metal-yielding
   b. industrial
   c. fuel
   * d. construction
   e. not

5. Lime and gravel are _____ minerals.
   a. metal-yielding
   * b. industrial
   c. fuel
   d. construction
   e. not

6. Schist is a _____ rock, formed by heat and pressure deep in the earth.
   a. sedimentary
   * b. metamorphic
   c. igneous
   d. bedrock
   e. surface

7. Developed countries have 25% of the world's population and consume ____% of the world's minerals.
   a. 25
   b. 50
   * c. 75
   d. 85
   e. 99

8. Aluminum and copper ore are _____ minerals.
   * a. metal-yielding
   b. industrial
   c. fuel
   d. construction
   e. not

9. Mineral consumption by the industrial nations has leveled off and in some cases declined. This is because these nations have largely completed their:
   a. automobile production.
   b. export of minerals to developing nations.
   c. stockpiling of minerals from developing nations.
   * d. infrastructure (roads, buildings, etc.)
   e. conquest of developing nations.

159

10. The metal mineral produced in the greatest amount worldwide is:
    a. aluminum
    b. copper
  * c. iron
    d. manganese
    e. zinc

11. The nonmetal mineral produced in the greatest amount worldwide is:
    a. salt
    b. sand
  * c. stone
    d. lime
    e. soda ash

12. The U.S. stockpiles which of the following minerals because they are *strategic minerals?*
  * a. bauxite
    b. lime
    c. copper
    d. lead
    e. zinc

13. The melting of a metal ore to extract the metal is called:
    a. tinning
    b. leaching
  * c. smelting
    d. separation
    e. chemical separation

14. This metal is mainly used to make the steel for automobiles and other vehicles:
    a. copper
    b. lead
  * c. iron
    d. magnesium
    e. silver

15. Gold is mined in the Amazon basin using high pressure streams of water; this type of mining is called:
    a. open pit
    b. strip mining
    c. underground
  * d. hydraulic
    e. surface

16. Gold is sometimes extracted by spraying a cyanide solution on piles of ore; this is technique is called:
    a. smelting
  * b. heap leaching
    c. stripping
    d. extraction
    e. fluxing

17. This process collects and returns used metals to factories where they are used again:
    a. restoration
  * b. recycling
    c. conservation
    d. positive feedback
    e. negative feedback

18. Manufacturing an aluminum can from recycled aluminum requires only ___% of the energy needed to make the can from raw ore.
  * a. 5
    b. 10
    c. 20
    d. 30
    e. 40

19. Approximately, ___% of all American cars are recycled.
    a. 20
    b. 40
    c. 50
    d. 70
  * e. 90

20. Copper and other nonferrous smelters produce about 8% of the world's _____ emssions.
    a. carbon dioxide
    b. nitrous oxide   d. ozone
  * c. sulfur dioxide   e. chlorine

21. Using only what we need and using it efficiently is called:
  * a. conservation
    b. recycling
    c. restoration
    d. preservation
    e. utility

22. The amount of time a metal or mineral stays in use is called the:
    a. rest period.
*   b. residence time.
    c. playing time.
    d. use time
    e. loop time.

23 The General Mining Act permits miners to purchase land for __ an acre or less.
*   a. $5
    b. $10
    c. $15
    d. $20
    e. $25

24. U.S. mining companies receive a tax exemption allowing them to deduct 5% to 22% of their gross earnings from their taxes each year. This is called the:
    a. coal allowance
    b. subsidy allowance
*   c. depletion allowance
    d. exploration allowance
    e. mineral allowance

25. The U.S. law that targets hazardous wastes is the:
    a. Clean Water Act
    b. Wilderness Act
*   c. Resource Conservation and Recovery Act
    d. General Mining Act
    e. Wild and Scenic Rivers Act

26. Deposits of minerals that are fairly certain to exist and are feasible to mine at current prices are called:
*   a. reserves
    b. surplus
    c. extractable
    d. new deposits
    e. undepleted reserve

27. About __% of mining in the U.S. occurs on private and state lands.
    a. 25
    b. 50
*   c. 75
    d. 80
    e. 95

28. According to the U.S. Bureau of Mines, hardrock mining has contaminated more than _____ miles of rivers and streams.
    a. 1,000
    b. 2,000
    c. 5,000
*   d. 10,000
    e. 20,000

29. According to the National Mining Association, the average American consumes _____ pounds of minerals each year.
    a. 10,000
    b. 20,000
    c. 30,000
*   d. 40,000
    e. 50,000

30. Mining lower grade ores produces greater environmental damage than mining higher-grade ores because:
    a. Larger mines are needed.
    b. More material will be transported to the smelters.
    c. More waste will be produced at the mines and smelters.
    d. More air and water pollution will result from increased mining activities.
*   e. all of the above

31. Manganese nodules have been discovered to be abundant at which of the following locations?
    a. Mars
    b. Moon
    c. Gobi Desert
*   d. Pacific Ocean
    e. Atlantic Ocean

32. In 1991 24 industrial nations signed a treaty banning environmentally damaging oil and mineral exploration in _____ for 50 years.
    a. Australia
*   b. Antarctica
    c. Polar regions
    d. Brazil
    e. Costa Rica

33. Buying durable products is a sustainable personal action encouraging:
    a. recycling
    *   b. conservation
    c. public/private support
    d. communication
    e. expensive products

34. Which of the following materials is a nonfuel mineral?
    *   a. zinc
    b. oil
    c. natural gas
    d. oil shale
    e. tar sands

35. *Patenting* is the process of:
    a. protecting your invention.
    *   b. filing a mining claim on a piece of federal land.
    c. ore extraction used in underground mining.
    d. exploring for new ore deposits.
    e. removing impurities from an ore.

36. The mining industry employs more than _____ thousand people in the U.S.
    a. 50
    b. 75
    c. 100
    *   d. 120
    e. 200

37. Which of the following factors reduce mineral supplies?
    a. High labor costs.
    b. Interest rates.
    c. Energy costs.
    d. Environmental protection costs.
    *   e. all of the above

38. One of the primary problems with mining the seafloor is:
    a. technical feasibility
    b. concentration of the ore deposit
    c. profitability of mining at sea
    *   d. ownership
    e. treaty violations

18

Creating Sustainable Cities and Towns: Principles and Practices of Sustainable Community Development

Chapter Outline

Cities and Towns as Networks of Systems
The Invisibility of Human Systems
Performance vs. Sustainability: Understanding a Crucial Difference
Why Are Human Systems Unsustainable?
The Challenge of Creating Sustainable Cities and Towns

Land-Use Planning and Sustainability
Sustainable Land-Use Planning: Choosing the Best Options
Statewide and Nationwide Sustainable Land-Use Planning
Beyond Zoning
Land-Use Planning in the Developing Nations

Shifting to a Sustainable Transportation System
Phase 1: The Move Toward Efficient Vehicles and Alternative Fuels
Phase 2: From Road to Rails, Buses, and Bicycles
Economic Changes Accompanying a Shift to Mass Transit

Key Terms

land-use planning	dispersed development	urban sprawl
compact development	satellite development	corridor development
greenbelts	zoning regulations	differential tax rate
development right	mass transit	telecommute

Objectives

1. List the human systems associated with cities and towns.
2. Discuss the two major challenges facing existing communities that are moving toward sustainability.
3. Define the following terms: "land-use planning", "dispersed development", "urban sprawl", "compact development", "satellite development", and "corridor development".
4. Compare the environmental impact of the community development approaches mentioned in question number three above.
5. Discuss what can be done by legislation to encourage sustainable land-use policies.
6. Define the following terms: "zoning regulations", "differential tax rate", and "development right".
7. Summarize what can be done to make our transportation systems more sustainable.
8. Discuss what you can do personally to make human systems more sustainable in the community.

Lecture Outline

I. Cities and Towns as Networks of Systems
    A. The Invisibility of Human Systems
        1. It is hard for most people to accept the fact that human systems are unsustainable
        2. Human systems are invisible to most people. Systems were designed to be unobtrusive to us in our everyday lives.
    B. Performance vs. Sustainability: Understanding a Crucial Difference
        1. Just because a system is functioning well does not necessarily mean that it is sustainable.
    C. Why Are Human Systems Unsustainable?
        1. Human systems exceed the carrying capacity of the earth.
    D. The Challenge of Creating Sustainable Cities and Towns
        1. The existing infrastructure must be revamped to incorporate the principles of sustainability: more efficient use of energy /resources. This is called the *redevelopment strategy*.
        2. New development must incorporate sustainable ideas and technologies.

II. Land-Use Planning and Sustainability
    A. Sustainable Land-Use Planning: Choosing the Best Options
        1. Dispersed Development (urban sprawl) - Dispersed development is the most environmentally unsustainable form of urban/suburban development.
            a. It consumes excessive amounts of land that is often prime farmland.
            b. Wildlife may be displaced from important habitat.
            c. Habitat destruction often increases flooding.
            d. The haphazard pattern of settlement produces esthetically unpleasant results.
            e. Vehicular traffic is increased adding to increased commuting time, energy consumption and air pollution.
            f. The costs of providing infrastructure generally costs more than in alternative forms of development.

2. Compact Development - This is the most efficient of the development alternatives. Compact development creates a clear line between city, a suburb and the outlying area.
   a. It requires less land than other forms of development.
   b. It helps to preserve open space, farmland, wildlife habitat, etc.
   c. Compact development minimizes vehicle miles traveled and can reduce energy consumption and air pollution.
   d. It is cheaper from an economic perspective than other forms of development.
   e. Compact development is successful in Europe.
B. Statewide and Nationwide Sustainable Land-Use Planning
   1. Land-use planning should be encouraged on a statewide and nationwide basis.
C. Beyond Zoning
   1. Many methods can help promote sustainable land-use patterns.
      a. Zoning Regulations
      b. Differential Tax Rates
      c. Development Right
      d. Making Growth Pay Its Own Way - Developers are held responsible for providing the cost of infrastructure in a new development project.
D. Land-Use Planning in the Developing Nations
   1. Land-use planning and land reform are also needed in the developing nations to encourage sustainable land-use and development.

III. Shifting to a Sustainable Transportation System - Transportation consumes 30% of the energy used in the United States.
A. Phase 1: The Move Toward Efficient Vehicles and Alternative Fuels
   1. Increased mileage in new cars and the development of autos which run on clean, alternative fuels can stretch our petroleum supplies and reduce transportation-related pollution.
B. Phase 2: From Road to Rails, Buses, and Bicycles
   1. Mass transit can replace much of the use of autos if made available to most people.
   2. Bicycles can be effective adjuncts to an efficient transportation system.
C. Economic Changes Accompanying a Shift to Mass Transit
   1. Transportation system changes will have major economic impacts; these can be minimized with proper planning

Suggestion for Presenting the Chapter

- Instructors should have students recognize the systems in their communities and evaluate their environmental impacts.
- A discussion of potential changes that would make local systems sustainable is an excellent activity. Discussion might focus on transportation: How is mass transit in your community? The focus should be on local issues which will affect your students.
- Examine the effects of development in your area. Is urban sprawl a problem? Is local government aware and responsive to issues of sustainability?
- A class activity might focus on a local system that can be improved to become more sustainable. An activism project focused on the issue will give students first hand experience.
- Recommended web sites:

165

Center of Excellence for Sustainable Development/Land Use Planning/U.S. Dept. of Energy:
http://www.sustainable.doe.gov/landuse/luintro.shtml
Sierra Club Challenge to Sprawl Campaign: http://www.sierraclub.org/sprawl/
Urban Environmental Management/Global Development Research Center:
http://www.gdrc.org/uem/
United Nations Sustainable Development: http://www.un.org/esa/sustdev/
Natural Resources Defense Council/Natural Resources and Green Living:
http://www.nrdc.org/cities/smartGrowth/default.asp

True/False Questions

1. ___ Human systems are unsustainable because they exceed the carrying capacity of the earth.
2. ___ Sustainable land-use planning seeks to use land more efficiently.
3. ___ Compact development is commonly referred to as urban sprawl.
4. ___ Satellite development involves the development of outlying communities connected to the metropolitan area by highways.
5. ___ Concentrating housing and business growth along major transportation corridors is called corridor development.
6. ___ The main tool of land-use planning for years has been greenbelt development.
7. ___ A development right is a law that classifies land according to use and prohibits certain land uses.
8. ___ More efficient vehicles and clean-burning fuels may be part of the first phase of the transition to a sustainable society
9. ___ Today, nearly 70% of the energy Americans consume is used by the transportation sector.
10. ___ The differential tax rate allows city officials to tax different lands at different tax rates.
11. ___ Mass transit is less efficient than using automobiles.
12. ___ Telecommuters take the commuter train to work.
13. ___ To be profitable, high-speed rail requires high-participation, high-density population in outlying areas and a large central business district.
14. ___ California is a catalyst for many of the changes occurring in automobile design.
15. ___ Greenbelts are undeveloped areas in or around cities and towns.
16. ___ Zoning can protect farmland and other lands from urban development.
17. ___ Urban sprawl is the sustainable, planned growth of residential areas.
18. ___ Mass transit uses buses, commuter trains and light rail to increase the efficiency of transportation.
19. ___ The most efficient mode of mass transportation is air travel.
20. ___ For decades, the bicycle has been a major means of transportation in many European and Asian countries.

Fill-in-the-Blank Questions

1. _____ , Texas has left yellow bicycles on the street for anyone who wants to use them.
2. The _____ industry is the world's largest manufacturing industry.
3. Today, ___ cents of every dollar spent in the United States is directly or indirectly connected to the automobile industry.
4. The employment potential of _____ transit exceeds that of the current automotive industry.
5. A _____ right is a fee paid to a farmer to prevent the land from being developed.
6. _____ regulations classify land according to use.
7. _____ development requires the least amount of land to accommodate people and the services they require.
8. _____ development concentrates housing and business growth along major transportation corridors.
9. _____ are undeveloped areas in or around cities and towns.
10. Urban _____ is the steady outward expansion of urban areas.
11. _____ plans set up industrial zones and residential zones.
12. If an energy system or manufacturing system appears to be functioning well does not mean it is _____ in the long run.
13. Revamping the existing infrastructure in cities is called _____.
14. Millions of hectares of _____ are destroyed each year by expanding urban centers.
15. The present systems are _____ because they produce waste and pollution in excess of the Earth's capacity to absorb them.
16. Urban sprawl is another name for _____ development.
17. Compact development not only minimizes land use, it also reduces vehicle _____ traveled.
18. The first step in the transition to a sustainable transportation system is an improvement in _____.
19. A personal action that would foster a more sustainable transportation system would be to drive an energy-_____ vehicle.
20. _____ and townhouses can accommodate far more people per hectare of land than single dwelling units.

Fill-in Key:

1. Austin
2. automobile
3. 20
4. mass
5. development
6. zoning
7. compact
8. corridor
9. greenbelts
10. sprawl
11. Land-use
12. sustainable
13. redevelopment
14. farmland
15. unsustainable
16. dispersed
17. miles
18. efficiency
19. efficient
20. Condominiums

Multiple Choice Questions

1. Just because a system such as energy is functioning well does not mean that it is

   _____.
   * a. sustainable
   b. efficient
   c. inefficient
   d. unsustainable
   e. good

2. For communities to become sustainable, they will need to:
   * a. Implement the redevelopment strategy.
   b. Continue conventional development.
   c. Eliminate existing development and start over.
   d. Ignore previous unsustainable systems.
   e. Abandon the current town and start over.

3. Redevelopment refers to:
   a. Proceeding with conventional development.
   * b. Refurbishing the existing infrastructure to be sustainable.
   c. Tearing down old structures and completely starting over.
   d. Planning for now, not the future.
   e. A sustainable cod e for new construction.

4. Retrofitting existing structures, such as a conventionally built home, with solar panels and a greenhouse is an example of:
   a. Green Revolution technologies.
   b. new development.
   * c. redevelopment.
   d. inefficient house construction.
   e. wasting money on frills.

168

5. Sustainable land use:
   a. Conserves farmland, and wildlife habitat.
   b. Encourages more efficient transportation.
   c. Reduces air and water pollution.
   d. Reduces the cost of providing water, sewage, and other forms of infrastructure.
   * e. all of the above

6. The most sustainable type of development is:
   a. dispersed development
   * b. compact development
   c. satellite development
   d. corridor development
   e. strip development

7. Urban sprawl is synonymous with this type of development:
   * a. dispersed development
   b. compact development
   c. satellite development
   d. corridor development
   e. strip development

8. This type of development involves outlying communities connected to the metropolitan area by highways:
   a. dispersed development
   b. compact development
   * c. satellite development
   d. corridor development
   e. strip development

9. This type of development concentrates housing and business growth along major transportation corridors:
   a. dispersed development
   b. compact development
   c. satellite development
   * d. corridor development
   e. highway development

10. A housing development with many condominiums, compact placement of services, a commons, and houses placed close to streets would be typical in _____ development.
    a. dispersed
    b. satellite
    * c. compact
    d. corridor
    e. strip

11. Undeveloped areas in and around cities are:
    a. croplands
    b. shelterbelts
    * c. greenbelts
    d. wasted space
    e. eyesores

12. National guidelines for land use have been established in which of the following countries?
    a. Belgium
    b. France
    c. the Netherlands
    d. Germany
    * e. all of the above

13. Laws which establish acceptable uses for land are called:
    a. development rights
    b. patents
    * c. zoning regulations
    d. development ordinances
    e. civil laws

14. A fee paid to a farmer to prevent land from being developed is a:
    a. retainer
    b. down payment
    c. development tax
    * d. development right
    e. capital investment

15. Each year Americans travel nearly _____ billion miles in their cars.
    a. 2
    b. 20
    c. 200
    * d. 2,000
    e. 20,000

169

16. The differential rate tax reduces urban sprawl because it:
    a. taxes businesses at a higher tax rate than housing developers
    b . discourages home owners to buy new houses through tax incentives.
    * c. taxes farmland at a lower rate than housing developments.
    d. taxes corporations at a higher rate than housing developments.
    e. relieves housing developers of their tax burden.

17. There are an estimated _____ million vehicles on the road in the United States.
    a. 20
    b. 80
    c. 120
    * d. 180
    e. 250

18. Today, nearly ___% of the energy Americans consume is used in transportation.
    a. 10
    b. 20
    * c. 30
    d. 40
    e. 50

19. The current transportation system is not sustainable because of:
    a. fossil fuel use.
    b. carbon dioxide release.
    c. acid deposition.
    d. exhaust pollutants.
    * e. all of the above

20. The average American automobile in 1995 got ___ miles to the gallon of gasoline.
    a. 15.5
    b. 20.5
    * c. 28.5
    d. 58
    e. 83

21. By the year 2000 all new cars sold in _____ will be ultraclean.
    a. Japan
    b. the Netherlands
    c. Oregon
    * d. Los Angeles
    e. Colorado

22. Buses, commuter trains, and light rail are forms of:
    a. rural infrastructure
    * b. mass transit
    c. obsolete transportation
    d. public transportation
    e. government socialism

23. Transportation by bus and train are about __ times more efficient than use of the average new car today.
    a. 2
    b. 3
    c. 5
    * d. 7
    e. 10

24. Employees that work at home linked to the office by phone lines are:
    a. home-based
    * b. telecommuting
    c. distance commuters
    d. non-commuters
    e. information linked

25. Which is the most inefficient type of transportation?
    a. rail
    b. bus
    c. car pool
    d. automobile
    * e. airline

26. Which of the following is a "hidden subsidy" of the automobile?
    a. traffic control
    b. gasoline subsidies
    c. city-paid parking facilities
    d. police protection
    * e. all of the above

27. Today about ___ cents of each dollar spent in the United States is directly or indirectly connected to the automobile.
    a. 10
  * b. 20
    c. 30
    d. 40
    e. 50

28. Which of the following is a responsible personal action for more sustainable transportation?
    a. walk more
    b. ride a bike
    c. take a bus or carpool
    d. drive an energy efficient vehicle
  * e. all of the above

29. Which of the following is a problem associated with urban sprawl?
    a. Loss of natural habitat.
    b. Increased chance of flooding.
    c. Destruction of prime farmland.
    d. Haphazard pattern of settlement with poor aesthetic appeal.
  * e. all of the above.

30. Electric cars:
    a. are sustainable alternative to the conventional cars.
    b. do not pollute since they use electricity from conventional power plants.
  * c. are almost as unsustainable as gas-powered cars if they use electricity from fossil fuel power plants.
    d. do worse than gas-powered cars on carbon monoxide and hydrocarbon pollution.
    e. must be used in California. by the year 2000.

171

19

Principles of Toxicology and Risk Assessment: Understanding How Pollution and Toxic Chemicals Affect Humans and Other Species

Chapter Outline

Principles of Toxicology
The Biological Effects of Toxicants
How Toxicants Work
Factors That Affect the Toxicity of Chemicals

Mutations, Cancer, and Birth Defects
Mutations
Cancer
Birth Defects

Reproductive Toxicity

Environmental Hormones

Case Studies: A Closer Look
Asbestos: How Great a Danger?
Electromagnetic Radiation: A Hazard to Our Health?

Controlling Toxic Substances: Toward a Sustainable Solution
Toxic Substances Control Act
Market Incentives to Control Toxic Chemicals
The Multimedia Approach to Pollution Control: An Integrated Approach

Risk and Risk Assessment
Risks and Hazards: Overlapping Boundaries
Three Steps in risk Assessment
Risk Acceptability

How Do We Decide If a Risk Is Acceptable?
The Final Filter: Ethics and Sustainability

Key Terms

pollution prevention	toxicology	toxicants
enzymes	mutations	hypersensitized
dose	dose-response curve	duration of exposure
acute	chronic	additive response
synergistic response	potentiation	bioaccumulation
biological magnification	mutations	mutagens
somatic cells	germ cells	cancer
benign tumor	malignant tumor	primary tumor
metastasis	secondary tumor	DNA-reactive carcinogens
epigenetic carcinogens	biotransformation	birth defects
teratogens	teratology	organogenesis
stillbirth	reproductive toxicity	environmental hormones
asbestosis	electromagnetic radiation	polychlorinated biphenyls
premanufacture notification	pollution taxes	green taxes
tradable permits	multimedia approach	anthropogenic hazards
natural hazards	risk assessment	hazard identification
estimation of risk	threshold level	risk acceptability
perceived harm	perceived benefit	cost-benefit analysis
space-time values		

Objectives

1. Define the following terms: "immediate toxicity", "delayed toxicity", "local effects" and "systemic effects".
2. List the three ways toxins are known to act at the cellular level.
3. List several factors that may affect the toxicity of chemicals.
4. Define the following terms: "dose", "dose-response curve", "LD_{50}", and "duration of exposure".
5. Give an example of biomagnification in the environment.
6. Define the following terms and give an example of each: "toxin", "mutagen", "carcinogen" and "teratogen".
7. Discuss the health effects of reproductive toxicants and environmental hormones.
8. Summarize the primary legislation regulating exposure to toxic substances in the United States.
9. Discuss some sustainable solutions to control the use of toxic substances.
10. Discuss the major sources of lead pollution, the health effects produced by exposure to lead and the ways in which human exposure can be reduced.
11. List the three steps used in risk assessment.
12. Discuss how we decide if a risk is acceptable.
13. Discuss the influence of ethics on our decisions concerning environmental risks.

Lecture Outline

I. Principles of Toxicology - Sixty thousand chemical substances are used commercially in the U.S. Of these, relatively few pose serious risks, yet those risks are substantial. Due to economic and time constraints, only a small percentage of potentially harmful chemicals are tested for sub-acute toxicity.

    A. The Biological Effects of Toxins

        1. Immediate and Delayed Toxicity

            a. Acute or immediate effects appear shortly often exposure and may be quite severe; they are usually the result of high-concentration short-term, exposures.

            b. Chronic effects are delayed responses generally due to long-term, low-level exposure.

        2. Reversible and Irreversible Toxic Effects

            a. While damage from exposure to some toxins is reversible, in other cases the damage is chronic or even lethal.

        3. Local and Systemic Toxic Effects

            a. Some toxins exert effects at only a localized site; others may effect an entire organ or organ system.

    B. How Toxins Work

        1. Toxins disrupt cellular metabolism in a variety of ways.

    C. Factors Affecting the Toxicity of Chemicals

        1. Dose and Duration of

            a. Higher doses and longer exposures tend to increase toxic effects.

            b. The LD_{50} is the dose of a toxin lethal to 50% of the test animals.

        2. Biological Activity and Route of Exposure

            a. Biological activity is a measure of a substance's reactivity in living systems.

            b. A toxin's route of entry determines how easily it enters the bloodstream and thus influences its effects.

        3. Age

            a. Younger organisms are usually more sensitive to toxins than are adults.

        4. Health Status

            a. Genetic predisposition and one's lifestyle may influence one's susceptibility to the effects of toxins.

        5. Chemical Interactions

            a. Two or more toxic substances present together can alter the expected response. A *synergistic response* is one in which the results are amplified; an *antagonistic response* is one in which the effects are negated; *potentiation* is the effect a nontoxic chemical has in intensifying the effect of a toxin.

        6. Bioaccumulation and Biological Magnification

            a. *Bioaccumulation* is the accumulation of certain chemicals in the body.

            b. *Biological magnification* is the buildup of chemicals through a food chain.

II. Mutations, Cancer, and Birth Defects

    A. Mutations

        1. Mutations are alterations of the genetic material.

        2. Agents which cause mutations are mutagens.

        3. Mutations of germ-cells can be passed from one generation to the next.

B. Cancer
   1. Cancer is an uncontrolled proliferation of cells, caused by one of a variety of agents or carcinogens.
   2. Environmental factors, food additives and other products, and occupational exposure account for most cancers not due to smoking.
C. Birth Defects
   1. These are physical biochemical or functional abnormalities.
   2. Agents which cause birth defects are teratogens; their effects depend on timing of exposure and the nature of the teratogen.

III. Reproductive Toxicity
   A. Many drugs and environmental agents can interfere with reproduction; the study of these is reproductive toxicology.

IV. Environmental Hormones
   A. *Environmental hormones* are chemical pollutants that enter organisms and can alter or mimic the bodies natural release of hormones. Hormones are regulatory chemicals produced by organs in the body.

V. Case Studies: A Closer Look
   A. Asbestos: How Great a Danger
      1. Asbestos is a naturally occurring element. It produces three serious disorders:
         a. pulmonary fibrosis
         b. lung cancer
         c. mesothelioma
      2. Asbestos is dangerous to asbestos workers and individuals who also smoke.
      3. The use of asbestos in the United States for insulation, fireproofing, and decorative purposes was banned in 1974. Further regulations enacted by the EPA in 1989 will eliminate almost 95% of all asbestos use in the United States.
   B. Electromagnetic Radiation: A Hazard to Our Health?
      1. Studies on the effects of extremely low-frequency (ELF) magnetic fields have been inconclusive up to this point in time.

VI. Controlling Toxic Substances: Toward a Sustainable Solution
   A. Federal Control
      1. The Toxic Substances Control Act is intended to screen and regulate chemicals to minimize human health risk.
      2. The Act requires premanufacture notification, testing, and hazard minimization.
   B. Market Incentives to Control Toxic Chemicals
      1. In California, a law requires manufacturers to either meet state-set standards for toxins or label their products accordingly, thus reducing the marketability of those products not in compliance.
   C. The Multimedia Approach to Pollution Control
      1. By training inspectors in several environmental media, the EPA hopes to control cross-media contamination.

VII. Risk and Risk Assessment
   A. Risks and Hazards: Overlapping Boundaries

1. Risk is a relative assessment of the threat posed by anthropogenic and natural hazards.
B. Three Step in Risk Assessment
1. Through hazard identification and risk estimation risk assessment seeks to help us understand and quantify risks in our lives.
2. Risk assessment is necessarily imprecise due to the nature of probability and severity estimates.
3. Due to the threshold level of many toxins, it is difficult to extrapolate from lab animals to humans in toxicity tests where high doses are used to speed and make observable any toxic effects.
C. Risk Acceptability
1. Since all activities involve some degree of risk, this is a relative term.
2. Risk acceptability is influenced by fear, perceived benefit, and perceived harm.
D. How Do We Decide If a Risk Is Acceptable?
1. Deciding which risks to accept in order to procure benefits involves tradeoffs.
2. Cost/benefit analysis is a popular decision-making tool though it is impossible to accurately quantify all benefits and costs relevant to most risk decisions.
3. Actual Versus Perceived Risk
a. When actual and perceived risk do not coincide, the result may be legislative over- or under protection.
E. The Final Filter: Ethics and Sustainability - Ultimately, our values determine our attitudes with regard to environmental risks.
1. Prioritizing Values
a. Our values must be prioritized in order for us to weigh and assess relative risks.
2. Space-Time Values
a. Our values also reflect certain spatial and temporal parameters which influence our decision-making.
3. Building a Sustainable Future
a. It is incumbent upon us to build a future in which risks are realistically assessed and minimized and where, if errors are made, they are made on the side of safety.

Suggestions for Presenting the Chapter

- Instructors should encourage the students to identify important environmental risks in their lives. How can we reduce or eliminate these risks? Are these risks associated with unsustainable systems/activities in our lives? How can these systems/activities be modified to become sustainable?
- A good exercise is to have students identify toxic chemicals in their homes. Students can be asked to research the harmful effects of the chemicals found and share their results in class or in a paper.
- Radon tests are easily performed and can be done in class at your educational institution. The results of these tests and the interpretation of those results provide a good review of the principles of toxicology. Students can be encouraged to perform radon tests in their own homes.

- Cancer is an excellent topic to explore with your classes. Today, much research is being done on the environmental aspects of our current epidemic. This is very relevant topic with students since few of us have not been affected in some way by this disease.
- Recommended web sites:

  U.S. Environmental Protection Agency/Pesticides and Toxics:
  	http://www.epa.gov/epahome/pestoxpgram.htm
  World Health Organization Pesticide Evaluation Scheme:
  	http://www.who.int/ctd/whopes/index.html
  Pesticide Action Network: http://www.panna.org/panna/
  U.S. Geological Survey Pesticide Use Maps: http://ca.water.usgs.gov/pnsp/use92/
  Animal and Plant Health Inspection Service/U.S. Dept. of Agriculture:
  	http://www.aphis.usda.gov/

True/False Questions

1. ___ Cellular proteins that regulate biochemical reactions are called enzymes.
2. ___ Potentially harmful changes in the structure of DNA of cells are known as biotransformation.
3. ___ The lethal dose of a toxin that kills half of the test animals is called the LD_{50}.
4. ___ Chemicals enter the body by three routes: inhalation, ingestion and dermal absorption.
5. ___ The duration of exposure is the amount of time that passes after initial exposure to a toxin.
6. ___ The fastest and most effective route of exposure is the skin.
7. ___ A response that is greater than the sum of the individual ones is known as a _____ response.
8. ___ Biological magnification refers to the increase in the concentration of a chemical substance within the food chain.
9. ___ DDT is a readily degradable insecticide and is not accumulated in body fat.
10. ___ When a nontoxic chemical combines with a toxic chemical and magnifies the toxicity of the latter, the effect is known as an additive response.
11. ___ Chemicals that cause mutations are called estrogens.
12. ___ Agents that cause birth defects are called teratogens.
13. ___ Chemical carcinogens that cause cancer without altering the DNA are called epigenetic carcinogens.
14. ___ The reproductive cells of organisms are known as germ cells.
15. ___ Metastasis is the accumulation of toxins in the tissues.
16. ___ The unchecked division of cells results in the formation of a tumor or neoplasm.
17. ___ Asbestos is a naturally occurring mercury compound.
18. ___ Asbestos is the only known cause of mesothelioma, a type of lung cancer.
19. ___ Skin and bone cells are somatic cells.
20. ___ Chronic exposures to a toxic substance last less than 24 hours.

Fill-in-the-Blank Questions

1. _____ cells are the reproductive cells of an organism.
2. _____ is only caused by exposure to asbestos fibers.
3. The amount of toxin that an animal is exposed to is known as the _____.
4. The ____ is the lethal dose of a toxin that kills half of the test animals.
5. The _____ of exposure is the amount of time an individual animal is exposed to a toxic substance.
6. Short term exposure to a toxin generally lasts less than ___ hours.
7. Chronic exposure to a toxin normally lasts more than __ months.
8. The slowest and least effective route of exposure of toxins is the _____.
9. _____ exposure is contact with a toxin by breathing.
10. Radioactive iodine released from nuclear accidents may be accumulated in the human _____ gland.
11. DDT is a persistent pesticide that may accumulate in body _____.
12. In biological _____ tissue concentrations of toxins increase at each level of the food chain.
13. _____ is the buildup of chemicals within the tissues and organs of the body.
14. _____ result from an uncontrolled growth of the tissues in the body.
15. The spread of cancerous cells in the body to other tissues is called _____.
16. Chemicals that cause mutations in the genetic material of an organism are known as _____.
17. Chemicals that cause birth effects are called _____.
18. Chemicals that increase the chance of development of cancer are known as _____.
19. The developmental period when organs are formed in the embryo is known as _____.
20. _____ carcinogens by cause cancer without altering the DNA directly.

178

11. fat
12. magnification
13. bioaccumulation
14. tumors
15. metastasis
16. mutagens
17. teratogens
18. carcinogens
19. organogenesis
20. Epigenetic

Multiple Choice Questions

1. Carbon monoxide, a pollutant in automobile exhaust binds to _____ preventing red blood cells from carrying oxygen.
    a. myoglobin
    * b. hemoglobin
    c. fibrin
    d. cholesterol
    e. renin

2. Toxic substances have there harmful actions by:
    a. binding to enzymes.
    b. interacting with the DNA.
    c. binding to nonenzyme molecules.
    d. a and b only
    * e. a, b, and c

3. The toxicity of a chemical is determined by the:
    a. route of entry.
    b. dose.
    c. length of exposure
    d. age of the individual
    * e. all of the above

4. The dose of a toxic substance that kills half of the test animals is called the:
    a. toxic dose
    b. threshold
    c. overdose
    * d. LD_{50}
    e. acute exposure

5. Acute or short-term exposure to a toxic substance are generally shorter than:
    a. 10 minutes
    b. 30 minutes
    c. 1 hour
    d. 12 hours
    * e. 24 hours

6. Chronic exposure to a chemical lasts more than:
    a. 1 week
    b. 1 month
    c. 2 months
    * d. 3 months
    e. 1 year

7. Which of the following responses are routes of exposure to chemicals?
    a. dermal exposure
    b. inhalation
    c. ingestion
    d. a and b only
    * e. a, b, and c

8. Ingestion refers to the process of:
    a. inhaling
    * b. eating
    c. injecting
    d. contact with the skin
    e. expectoration

9. The combination of barbiturate tranquilizers and alcohol are an example of:
    a. additive effect
    *b. synergistic effect
    c. potentiation
    d. antagonism
    e. bioaccumulation

10. When a chemical with no toxic effect combines with a toxic chemical and makes the toxin more potent this effect is:
    a. additive
    b. synergistic
    *c. potentiation
    d. antagonism
    e. bioaccumulation

11. When chemicals negate each other's effect in the body the process is called:
    a. additive
    b. synergistic
    c. potentiation
    *d. antagonism
    e. bioaccumulation

12. The buildup of chemicals within tissues of the body is called:
    *a. bioaccumulation
    b. additive effect
    c. potentiation
    d. synergism
    e. toxicity

13. The progressive increase in the concentration of a chemical substance in the organisms of a food chain is called:
    a. additive effect
    b. synergistic effect
    *c. biological magnification
    d. potentiation
    e. antagonism

14. This persistent pesticide is well know for its devastating effects on bird populations:
    a. DDE
    *b. DDT
    c. malathion
    d. parathion
    e. dieldrin

15. Changes in the genetic material are called:
    a. adaptations
    b. alterations
    c. hybrids
    *d. mutations
    e. variations

16. Mutation refers to:
    a. Changes in the DNA molecule.
    b. Alterations of the chromosome that are visible or detectable.
    c. Missing or extra chromosomes.
    d. a or b only
    *e. a, b, or c

17. The cells of the reproductive organs are called:
    a. somatic cells
    b. stroma
    c. connective tissue
    *d. germ cells
    e. parenchyma

18. Genetic mutations are present in about ___ of every 100 newborns.
    a. 1
    *b. 2
    c. 5
    d. 10
    e. 50

19. Down Syndrome is caused by:
    a. changes in the DNA of a chromosome.
    b. alterations in an enzyme.
    *c. abnormal chromosome numbers.
    d. a deformed chromosome.
    e. a defective hormone.

20. In the United States alone, cancer kills _____ people annually.
    a. 5,000
    b. 10,000
    c. 25,000
    d. 100,000
    *e. 500,000

21. Cancers occur most commonly in _____ tissue.
    a. lung
    b. muscle
    c. skin
    d. a and b only
*   e. a and c only

22. Breast cancer tends to _____ to bone.
    a. proliferate
    b. agglutinate
*   c. metastasize
    d. migrate
    e. metamorphose

23. Chemicals that cause cancer are called:
    a. antibiotics
    b. hazardous
*   c. carcinogens
    d. mutagens
    e. teratogens

24. Chemicals that induce cancer by altering the DNA of cells are called:
    a. toxic chemicals
*   b. genotoxic carcinogens
    c. epigenetic carcinogens
    d. teratogens
    e. ecoestrogens

25. Estrogens are sex hormones that stimulate:
    a. aging.
*   b. cell division.
    c. male development.
    d. hair loss.
    e. cell death.

26. Chemical carcinogens that induce cancer without altering DNA directly are called:
    a. DNA-reactive carcinogens
*   b. epigenetic carcinogens
    c. teratogens
    d. toxic chemicals
    e. hazardous chemicals

27. Chemical carcinogens are chemically altered in the body, a process called:
    a. bioaccumulation
    b. biomagnification
*   c. biotransformation
    d. mutation
    e. epigenesis

28. About __% of all newborns have a birth defect.
    a. 0.7
    b. 1.0
    c. 3.5
*   d. 7.0
    e. 15.0

29. Agents that cause birth defects are called:
    a. genotoxic carcinogens
    b. epigenetic carcinogens
*   c. teratogens
    d. toxic chemicals
    e. hazardous chemicals

30. The study of birth defects is called:
    a. embryology
    b. developmental biology
    c. pediatrics
*   d. teratology
    e. oncology

31. Which of the following factors are known to cause birth defects?
    a. progesterone
    b. thalidomide
    c. German measles
    d. alcohol
*   e. all of the above

32. At high levels of exposure dioxins are thought to be carcinogenic; at low levels they may suppress the immune system and may be a/an:
    a. insecticide.
    b. tumor suppressant.
*   c. environmental hormone.
    d. hazardous chemical.
    e. teratogen.

33. This substance was in widespread use for insulation, a soundproofing agent, and is found in auto brake pads, brake linings and other goods. It produces severe lung damage, this substance is:
    a. lime
    b. iron
    c. aluminum
* d. asbestos
    e. latex

34. Between 1990 and 2020, some health experts predict about __ million people will die from exposure to asbestos.
    a. 0.5
    b. 1
* c. 2
    d. 3
    e. 4

35 This law requires premanufacture notification by chemical companies before they import or manufacture a new chemical:
* a. Toxic Substance Control Act
    b. Resource Conservation and Recovery Act
    c. Clean Air Act
    d. Clean Water Act
    e. Wilderness Act

36. Which of the following responses is a source of lead pollution?
    a. power plants
    b. smelters
    c. automobile exhaust
    d. boilers that burn used motor oil
* e. all of the above

37. Fees levied on environmentally unsound activities are called:
    a. luxury taxes
    b. use taxes
* c. green taxes
    d. excise taxes
    e. regressive taxes

38. One of the reasons children are more susceptible than adults to lead poisoning is::
    a. Children do not eat substances which remove lead from their bodies.
* b. Children absorb more lead in their digestive tracts than adults.
    c. Children are less likely to ingest lead-based paints.
    d. Adults come in contact with car exhaust fumes on a regular basis.
    e. Lead has been eliminated in the environment of adults.

39. A multimedia approach to regulate toxic chemicals:
* a. Focuses on training inspectors in more than one type of pollution to curb abuses in disposal of toxic wastes.
    b. Uses radio and TV commercials to bring the message to the masses.
    c. Looks at different chemical types and their effects on the environment.
    d. Uses private and state monitoring systems.
    e. Lets industry monitor their own wastes.

40. Hazards created by human beings are called _____ hazards.
    a. human
    b. natural
* c. anthropogenic
    d. misanthropic
    e. primary

20

Air Pollution and Noise: Living and Working in a Healthy Environment

Chapter Outline

Air: The Endangered Global Commons
    Sources of Air Pollution
    Anthropogenic Air Pollutants and Their Sources
    Primary and Secondary Pollutants
    Toxic Air Pollutants
    Industrial and Photochemical Smog
    Air Pollution - A Symptom of Unsustainable Systems?

The Effects of Climate and Topography on Air Pollution
    The Cleansing Effects of Wind and Rain: Don't Be Fooled
    Mountains and Hills
    Temperature Inversions

The Effects of Air Pollution
    The Health Effects of Air Pollution
    Effects on Plants and Nonhuman Animals
    Effects on Materials

Air Pollution Control: Toward a Sustainable Strategy
    Cleaner Air Through Better Laws
    Cleaner Air Through Technology: End-of-Pipe Solutions
    Economics and Air Pollution Control
    Toward a Sustainable Strategy

Noise: The Forgotten Pollutant
    What is Sound?
    What is Noise?
    Impacts of Noise
    Controlling Noise

Indoor Air Pollution
    How Serious Is Indoor Air Pollution?
    Controlling Indoor Air Pollutants

Key Terms

anthropogenic pollutants	nonanthropogenic pollutants	criteria air pollutants
primary air pollutants	secondary air pollutants	toxic air pollutants
gray-air cities	brown-air cities	photochemical smog
cross-media contamination	temperature inversion	subsidence inversion
radiation inversion	chronic bronchitis	emphysema
bronchial asthma	ozone	sulfur dioxide
sulfuric acid	nitric acid	nitrogen oxide
attainment area	nonattainment area	pollution taxes
noncompliance penalties	end-of-pipe solutions	catalytic converters
scrubber	electrostatic precipitator	cyclones
magnetohydrodynamics	fluidized bed combustion	decibels
pitch	hertz	noise
indoor air pollutants	sick building syndrome	radon

Objectives

1. List the sources and environmental impacts of anthropogenic air pollution.
2. Define the following terms: "criteria air pollutants", "primary pollutants", "secondary pollutants" and "toxic air pollutants".
4. List the six major criteria air pollutants, their sources and potential public health effects.
5. Define the terms: "gray-air cities", "brown-air cities", and "photochemical smog".
6. Discuss the effects of climate and topography on air pollution.
7. List most common acute and chronic health effects of air pollution.
8. Summarize the major provisions of the Clean Air Act and its amendments.
9. Define the following terms: "nonattainment area", "emissions offset policy", "attainment region", "noncompliance penalties", and "pollution taxes".
10. Summarize the current "end-of-pipe" technologies for air pollution control.
11. List some pollution prevention strategies that provide a sustainable approach to air quality.
12. Define the following terms: "sound", "noise", "decibel", "pitch", "frequency", and "hertz".
13. Summarize the major laws regulating noise in the United States.
14. List some ways noise pollution can be controlled in the environment.
15. Identify the major types and sources of indoor air pollution.

Lecture Outline

I. Air: The Endangered Global Commons
  A. Sources of Air Pollution
    1. Natural and Anthropogenic Sources
      a. Though much air pollution is natural in origin, anthropogenic pollution poses the biggest environmental threat.
  B. Anthropogenic Air Pollutants and Their Sources
    1. Combustion is the single largest cause of air pollution today.
    2. Incomplete combustion and the presence of mineral contaminants in fuels generate a variety of harmful pollutants.

C. Primary and Secondary Pollutants
   1. Reactions between atmospheric primary pollutants produce a variety of secondary pollutants.
D. Toxic Air Pollutants
   1. Hundreds of toxic pollutants are released into the air, the U.S. has not regulated these in the past.
E. Industrial and Photochemical Smog
   1. Gray-Air and Brown-Air Cities
      a. Gray-air cities are usually characterized by cold moist climates; the major pollutants are sulfur oxides and particulates.
      b. Brown-air cities are typically in warn, dry, sunny climates; the major pollutants are carbon monoxide, hydrocarbons, and nitrogen oxides; these cities are subject to enshrouding by photochemical smog.
F. Air Pollution - A Symptom of Unsustainable Systems?
   1. Air pollution like other forms of environmental deterioration is a symptom of unsustainable systems of transportation, industry, housing, and energy production.

II. The Effects of Climate and Topography on Air Pollution
   A. The Cleansing Effects of Wind and Rain: Don't Be Fooled
      1. Factors Affecting Air Pollution Levels
         a. Wind and Rain
            i. These agents can clear the air but may result in cross-media contamination.
         b. Mountains and Hills
            i. These geographic features can trap pollutants in an area, inhibiting dispersion
         c. Temperature Inversions
            i. These are inverted temperature profiles, either due to movement of large air masses or uneven daily cooling of air and ground, which trap pollutants and allow them to accumulate over an area.

III. The Effects of Air Pollution
   A. Health Effects
      1. Immediate Health Effects
         a. Immediate effects of air pollution are those experienced immediately after exposure.
      2. Chronic Health Effects
         a. Chronic effects on human health result from long-term exposure to air pollution and include several serious diseases such as emphysema, bronchitis, and lung cancer.
      3. High-Risk Populations
         a. Some groups, such as the very young, the very old and the infirm, are especially susceptible to air pollution's effects.
   B. Effects on Plants and Nonhuman Animals
      1. Livestock, wildlife, and wild and cultivated plants can all be adversely affected by air pollution.
   C. Effects on Materials

1. Air pollution causes severe damage to buildings, most materials, and many priceless works of art.

IV. Air Pollution Control: Toward a Sustainable Strategy
   A. Cleaner Air Through Better Laws
      1. The Clean Air Act with its amendments is the major legal instrument for air quality protection in the U.S.
      2. The CAA's provisions aim to protect air that is already clean and promote improvement in areas already polluted; enforcement is the responsibility of the EPA. Recent revisions of the CAA address acid precipitation, toxics, automobile fuel efficiency standards, and ozone depletion; in addition, they provide market incentives for companies to reduce their pollution.
   B. Cleaner Air Through Technology
      1. Stationary Sources
         a. Reductions in air pollution from stationary sources are usually achieved by the removal of pollutants from emissions gases.
         b. Filters, cyclones, precipitators, and scrubbers remove particulates and most polluting gases but generate large amounts of hazardous waste.
      2. Mobile Sources
         a. Emissions from vehicles are generally controlled by conversion to nonpolluting substances.
      3. New Combustion Technologies
         a. New combustion techniques such as magnetohydrodynamics, fluidized bed combustion, and gas turbines increase efficiency and reduce air pollution from burning coal and natural gas.
   C. Economics and Air Pollution Control
      1. Though costly, numerous studies estimate the benefits of air pollution control measures to exceed their costs.
   D. Toward a Sustainable Strategy
      1. Though pollution control is necessary, a truly sustainable approach to air quality will emphasize conservation, recycling, and renewable energy.
      2. Individuals as well as corporations and governments must take responsibility for minimizing resource demand and pollution.

V. Noise: The Forgotten Pollutant
   A. What is Sound?
      1. Sound waves are compression waves that travel through the air.
      2. Sound is characterized by loudness (measured in decibels) and pitch (high or low the frequency is).
   B. What is Noise?
      1. Noise is an unwanted, unpleasant sound.
   C. Impacts of Noise
      1. Noise affects us in many ways. It damages hearing, disrupts our sleep, and annoys us in our everyday lives.
   D. Controlling Noise
      1. Noise levels can be controlled by redesigning machinery and other noise sources. Buildings can be sound insulated and noise generators can be separated from people by other measures.

VI. Indoor Air Pollution
    A. How Serious Is Indoor Air Pollution?
        1. Indoor air pollutants may actually be the cause of tens of thousands of deaths and much illness in industrialized nations.
        2. Major indoor air pollutants are:
            a. tobacco smoke
            b. radon
            c. formaldehyde
            d. asbestos
    B. Controlling Indoor Air Pollutants
        1. Many behavioral, regulatory, and technological options are available to reduce our exposure to indoor air pollutants.

Suggestions for Presenting the Chapter

- Instructors should review the impact of the Clean Air Act and its amendments. Information about the air quality in your region is available from the EPA.
- Instructors should emphasize that most of our strategies to reduce air pollution depend on end-of-pipe technologies. Pollution prevention needs to be emphasized to move to a more sustainable society.
- A field trip to a local power plant or large industry to examine their air pollution control technologies provides students with insights relevant to the learning objectives of this chapter.
- Radon levels might be measured at your educational institution. Kits are relatively inexpensive and provide interesting discussion material
- Recommended web sites:

    U.S. EPA/Air Emissions: http://www.epa.gov/ebtpages/airairpoemissions.html
    U.S. EPA/Plain English Guide to the Clean Air Act:
                      http://www.epa.gov/oar/oaqps/peg_caa/pegcaain.html
    U.S. EPA/Airnow - Real Time Air Pollution Data: http://www.epa.gov/airnow/
    U.S. EPA/Radon: http://www.epa.gov/iaq/radon/
    Sierra Club Clean Air Program: http://www.sierraclub.org/cleanair/

True/False Questions

1. ___ Nonanthropogenic pollutants come from human sources.
2. ___ Air contains 24% nitrogen and 78% oxygen.
3. ___ Carbon dioxide is one of the criteria air pollutants.
4. ___ Fossil fuels consist primarily of carbon and hydrogen.
5. ___ Pollutants released into the atmosphere are called primary pollutants.
6. ___ Acid rain formation would be an example of a secondary pollutant.
7. ___ Denver, Los Angeles and Salt Lake City are brown-air cities.

8. ___ Older industrial cities with moister climates, particulate emissions and sulfur oxides are brown-air cities.
9. ___ Air pollutants that end up contaminating water or land on earth is an example of cross-media contamination.
10. ___ Mountains and hilly terrain can impede the flow of air resulting in the buildup of pollutants in cities and industrialized areas.
11. ___ Temperature inversions trap warm air near the earth beneath a layer of very cold air.
12. ___ A subsidence inversion occurs when a high pressure air mass slides over a colder air mass.
13. ___ Air pollution damages many human-made materials from metal to concrete.
14. ___ The Federal Clean Air Act was passed by the Congress in 1963.
15. ___ Attainment regions are regions where air quality meets federal standards.
16. ___ Catalytic converts transform ozone into oxygen and carbon dioxide.
17. ___ Sound waves are compression waves that travel through the air.
18. ___ Loudness is measured in amperes.
19. ___ Cycles per second are commonly called hertz.
20. ___ Noise is an unwanted, unpleasant sound.

True/False Key:

1. F 2. F 3. F 4. T 5. T 6. T 7. T 8. F 9. T 10. T 11. F 12. T 13. T 14. T 15. T 16. F 17. T 18. F 19. T 20. T

Fill-in-the-Blank Questions

1. The U.S. EPA identifies four indoor air pollutants as the most dangerous: cigarette smoke, formaldehyde, asbestos and _____.
2. A permanent _____ shift occurs after continued exposure to loud noise.
3. _____ is an unwanted, unpleasant sound.
4. OSHA's workplace standard is __dB for an eight-hour exposure.
5. Loudness is measured in _____.
6. Fluidized bed combustion is a technology associated with _____-fired burners.
7. Pollution _____ is a key element of a sustainable society.
8. Smokestack scrubbers and electrostatic precipitators are examples of ___-__-____ controls.
9. _____ convertors transform carbon monoxide and hydrocarbons into water and carbon dioxide.
10. A _____ removes particulates and gases from air by using a mist of water and lime.
11. The Clean Air Act was passed by Congress in _____.
12. _____, sulfur dioxide and sulfuric acid are the pollutants most hazardous to plants.
13. Chronic _____ is a persistent inflammation of the bronchial tubes.
14. At least 2000 people died in the country of _____ as a result of the release of 40 tons of methyl isocyante into the atmosphere.

15. Temperature _____ are warm-air lids that form over cities trapping polluted air for days or weeks near the ground.
16. Air pollutants in air can contaminate waterways or soils, this pollution is known as ____-____ contamination
17. _____ smog forms when pollutants in the air react in the presence of sunlight.
18. Sulfuric acid formed in the atmosphere, from the reaction of sulfur dioxide with water, is a _____ pollutant.
19. _____ dioxide can react in the atmosphere to form nitric acid and is a key reactant in the formation of photochemical smog.
20. Anthropogenic air pollutants come from _____, energy production and industry.

Fill-in Key:

1. radon
2. threshold
3. noise
4. 90
5. decibels
6. coal
7. prevention
8. end-of-pipe
9. catalytic
10. scrubber
11. 1963
12. ozone
13. bronchitis
14. India
15. inversions
16. cross-media
17. photochemical
18. secondary
19. nitrogen
20. transportation

Multiple Choice Questions

1. The most abundant gas in air is:
    a. hydrogen
    b. oxygen
    *   c. nitrogen
    d. helium
    e. carbon dioxide

2. Which of the following would be an air pollutant from a forest fire?
    a. carbon monoxide
    b. carbon dioxide
    c. nitrogen oxides
    d. particulates
    *   e. all of the above

3. Anthropogenic air pollutants come from which of the following sources?
    a. volcanoes
    b. forest fires
    c. soil
    *   d. energy production
    e. plants

4. Carbon dioxide is a:
    *   a. primary pollutant
    b. secondary pollutant
    c. toxic air pollutant
    d. solely an anthropogenic pollutant
    e. hazard because of flammability.

5. Photochemical smog is a:
    a. primary pollutant
    *   b. secondary pollutant
    c. a natural pollutant
    d. chemical formed by the combination of water and carbon dioxide
    e. pollutant that only forms at night

6. Which of the following substances is a mineral contaminant found in fossil fuels?
    a. lime
    b. benzene
    *   c. lead
    d. nitrogen gas
    e. water

7. The major source of hydrocarbon pollution in air is:
    a. industrial combustion
    b. stationary combustion
    *   c. transportation
    d. mining
    e. smelters

8. Nitrogen dioxide forms when nitrogen combines with _____ at high temperatures.
    a. water
    b. oil
    *   c. oxygen
    d. sulfur
    e. carbon

9. Industrial cities, like London and New York are _____ cities.
    a. brown-air
    *   b. gray-air
    c. black-air
    d. yellow-air
    e. fog

10. Denver, Los Angeles, and Albuquerque are _____ cities.
    *   a. brown-air
    b. gray-air
    c. black-air
    d. yellow-air
    e. smog

11. This oxygen containing molecule is one of the most prevalent chemicals in photochemical smog and is involved in absorption of ultraviolet radiation:
    a. carbon dioxide
    b. nitrous oxide
    *   c. ozone
    d. sulfuric acid
    e. nitric acid

12. The transfer of airborne pollutants to surface
    waters is called:
    a. rain
    * b. cross-media contamination
    c. surface contamination
    d. water pollution
    e. gray-air contamination

13. When a warm air mass traps cooler air near
    the ground there is a:
    a. radiation inversion
    * b. subsidence inversion
    c. ozone layer
    d. density inversion
    e. normal pattern

14. The American Lung Association estimates
    that air pollution costs Americans ___ billion
    a year in health costs.
    a. $10
    b. $20
    c. $30
    d. $40
    * e. $50

15. Long term exposure to air pollution may
    result in a number of diseases including:
    a. bronchitis
    b. emphysema
    c. lung cancer
    d. asthma
    * e. all of the above

16. One out of every five American men between
    the ages of 40 and 60 suffers from chronic:
    a. pneumonia
    b. emphysema
    * c. bronchitis
    d. pleurisy
    e. asthma

17. During an attack of this disease the
    passageways that carry air to the lungs fill
    with mucus and narrow making it difficult to
    breath:
    a. pleurisy
    * b. asthma
    c. chronic rhinitis
    d. conjunctivitis
    e. emphysema

18. This group of the population is highly
    susceptible to air pollution:
    a. the young.
    b. the old.
    c. the infirm.
    d. a and b only
    * e. a, b, and c

19. Ozone, sulfur dioxide and sulfuric acid are
    pollutants that are very hazardous to:
    a. bridges
    b. humans
    * c. plants
    d. bacteria
    e. viruses

20. Air quality and the spread of pollution from
    an industrial source can be determined by
    mapping the presence or absence of this
    group of organisms:
    a. molluscs
    b. birds
    * c. lichens
    d. butterflies
    e. voles

21. The Taj Mahal in India is being defaced by
    _____ released from local power plants.
    a. carbon dioxide
    b. ozone
    * c. sulfur dioxide
    d. fly ash
    e. heavy metals

22. Ozone is particularly damaging to:
    a. rock
    b. plastic
    c. concrete
    * d. rubber
    e. steel

23. The "end-of-pipe" strategy for controlling
    pollution works by:
    a. capturing pollutants
    b. converting pollutants to less harmful
       substances
    c. reducing activities that produce pollution
    * d. a and b only
    e. b and c only

191

24. The primary piece of legislation regulating air pollution in the U.S. is the:
*   a. Clean Air Act
    b. Resource Conservation and Recovery Act
    c. Clean Water Act
    d. Federal Hazardous Substances Act
    e. Toxic Substances Control Act

25. The ambient air quality standards established by the EPA cover:
    a. toxic air pollutants.
    b. carcinogens in the air.
*   c. six "criteria" air pollutants.
    d. pollution taxes.
    e. tradable permits for polluters.

26. Areas that are violating the ambient air quality standards are called:
    a. attainment regions
*   b. nonattainment regions
    c. subsidence regions
    d. noncompliance regions
    e. hazardous areas

27. The rules for the prevention of significant deterioration (PSD) of air quality in areas meeting the ambient air quality standards are part of the:
    a. Clean Air Act of 1970
*   b. 1977 amendments to the Clean Air Act
    c. 1990 amendments to the Clean Air Act
    d. Toxic Substances Control Act
    e. Resource Conservation and Recovery Act

28. Noncompliance penalties were able to be assessed polluters under the:
    a. Clean Air Act of 1970
*   b. 1977 amendments to the Clean Air Act
    c. 1990 amendments to the Clean Air Act
    d. Toxic Substances Control Act
    e. Resource Conservation and Recovery Act

29. Pollution taxes, a market based incentive, for reducing toxic chemical emissions was established by the:
    a. Clean Air Act of 1970
    b. 1977 amendments to the Clean Air Act
*   c. 1990 amendments to the Clean Air Act
    d. Toxic Substances Control Act
    e. Resource Conservation and Recovery Act

30. Tradable permits are available to companies and can be sold at a profit if unused. This provision was established by the:
    a. Clean Air Act of 1970
    b. 1977 amendments to the Clean Air Act
*   c. 1990 amendments to the Clean Air Act
    d, Toxic Substances Control Act
    e. Resource Conservation and Recovery Act

31. The production of ozone-depleting chemicals is phased out in the U.S. by the:
    a. Clean Air Act of 1970
    b. 1977 amendments to the Clean Air Act
*   c. 1990 amendments to the Clean Air Act
    d. Toxic Substances Control Act
    e. Resource Conservation and Recovery Act

32. Bag filters, cyclones, and electrostatic precipitators are all:
    a. devices to prevent pollutants from being produced.
*   b. end-of-pipe strategies.
    c. catalytic converters.
    d. pollution prevention strategies.
    e. state of the art technologies.

33. This pollution control technology removes sulfur dioxide and particulates. Pollutant laden air is passed through a fine mist of water and lime which traps the pollutants; this technology is a:
    a. electrostatic precipitator
*   b. scrubber
    c. cyclone
    d. bag filter
    e. catalytic converter

34. This pollution control device transforms carbon monoxide and hydrocarbons into water and carbon dioxide:
   a. bag filter
   b. electrostatic precipitator
   c. scrubber
 * d. catalytic converter
   e. cyclone

35. This pollution control device uses centrifugal force to remove particulates from air passed through a metal cylinder:
   a. bag filter
   b. electrostatic precipitator
   c. scrubber
   d. catalytic converter
 * e. cyclone

36. Coal is mixed with an ion-producing seed substance, such as potassium and burned in this efficient technology:
   a. fluidized bed combustion
 * b. magnetohydrodynamics
   c. convection combustion
   d. injection combustion
   e. plasma arc combustion

37. Loudness is measured in:
   a. joules
   b. newtons
   c. kilograms
 * d. decibels
   e. picocuries

38. Noise in the U.S. workplace is controlled by the:
   a. Noise Control Act
 * b. Occupational Safety and Health Administration
   c. Environmental Protection Agency
   d. Department of Defense
   e. U.S. Department of Agriculture

39. This legislation allowed the EPA to establish maximum permissible noise levels for motor vehicles and other sources:
 * a. Noise Control Act
   b. Consumer Product Safety Act
   c. Occupational Health and Safety Act
   d. Dangerous Cargo Act
   e. Resource Conservation and Recovery Act

40. Which of the following substances are classified by the EPA as dangerous indoor air pollutants?
   a. cigarette smoke
   b. radon
   c. formaldehyde
   d. asbestos
 * e. all of the above

21

Global Air Pollution: Ozone Depletion, Acid Deposition, and Global Warming

Chapter Outline

Stratospheric Ozone Depletion

Acid Deposition: Ending the Assault

Global Warming/Global Climate Change

Key Terms

ozone layer   chlorofluorocarbons   ultraviolet radiation
acid deposition   pH   acid
wet deposition   acid precursors   dry deposition
buffers   buffering capacity   greenhouse gases
water vapor   carbon dioxide   nitrous oxide
methane   greenhouse effect   global warming
global climate change

Objectives

1. List the major activities which deplete ozone and discuss the extent, effects, and prevention of ozone depletion.
2. List the substitutes for ozone-destroying chlorofluorocarbons.
3. Discuss the sources and transport of acid precursors.
4. Summarize the major environmental impacts of acid deposition.
5. List the short-term and long-term solutions to acid deposition.
6. List the major greenhouse gases, their relative contribution to the greenhouse effect and principle sources.
7. Discuss some of the potential ecological impacts of global climate change due to global warming.
8. Discuss how application of the principles of sustainability could reduce the problem of global warming.

Lecture Outline

    I.   Stratospheric Ozone Depletion - The ozone layer in the stratosphere protects life on earth from harmful levels of ultraviolet radiation.
        A. Activities That Deplete the Ozone Layer
            1. Freons
               a. Freons, or CFCs, are used as spray propellants, blowing agents, and coolants break down ozone at an alarming rate.
            2. High-Altitude Jets
               a. High speed, high altitude air- and spacecraft emit nitric oxides, which destroy ozone.
        B. Ozone Depletion - The History of a Scientific Discovery
            1. Studies of the ozone layer show substantial declines over the globe, with the highest level of depletion in the southern hemisphere and Antarctica.
        C. The Many Effects of Ozone Depletion
            1. Ozone depletion could seriously affect human health, ecosystems, crops, and materials and finishes.
        D. Banning CFCs and Other Ozone Depleting Chemicals: A Global Success Story

1. Nations of the world have responded to the threat of ozone depletion; three international treaties have already been signed to eliminate the production of ozone-depleting chemicals.

E. Substitutes for Ozone-Destroying CFCs
   1. CFCs are being replaced by HCFCs because HCFCs are much less harmful to the ozone layer.
   2. Research continues to find ozone-friendly substitutes.

F. The Good News and Bad News About Ozone
   1. CFCs take 15 years or so to migrate into the stratosphere after release.
   2. It will take at least 100 years before the ozone levels return to 1985 levels.
   3. Another 100-200 years may be needed to have a complete recovery of the ozone levels in the stratosphere.

II. Acid Deposition: Ending the Assault - This phenomenon constitutes one of the most serious environmental and economic threats facing us today.
   A. What Is an Acid?
      1. Acids are chemical substances that add hydrogen ions to a solution.
   B. What Is Acid Deposition?
      1. Acid deposition is the deposition of mostly sulfuric and nitric acids from the sky on soils and in bodies of water.
         a. Wet Deposition
            i. Rain, snow, fog, or clouds may deposit acid in wet form.
         b. Dry Deposition
            i. Sulfur and nitrogen oxides may settle directly out of the air or may form particulates which, on surfaces, combine with water to form acids.
   C. Where Do Acids Come From?
      1. Sources for acid precursors are natural and anthropogenic.
      2. Volcanoes, forest fires, and bacterial decay are natural sources of sulfur oxides.
      3. Anthropogenic sources include fossil-fuel fired power-plants, motor vehicles, and ore smelters.
   D. The Transport of Acid Precursors
      1. Acid precursors generated in industrial regions may travel hundreds or thousands of kilometers downwind.
   E. The Social, Economic, and Environmental Impacts of Acid Precipitation
      1. Acidification of Lakes
         a. Thousands of lakes and streams in the eastern U.S., Canada, and Scandinavia are alarmingly acidified; fish and other aquatic organisms are severely impacted.
      2. Widening the Circle of Destruction
         a. Acid deposition affects both aquatic species and all species that depend on the aquatic ecosystem for food.
      3. Forest and Crop Damage
         a. Acid deposition damages forests and crops in many parts of the world. Plants are damaged directly by acids but also indirectly through changes in soil chemistry and soil-dwelling organisms.
      4. Acids: Fertilizing Effect

        a. The sulfur and nitrogen released by the acids in the soil can promote plant growth. Damaging effects of the acid conditions usually outweigh the benefits resulting from their fertilizing effect.

    5. Damage to Materials

        a. Many manmade materials, including culturally valuable artifacts, suffer irreparable or costly damage as a result of acid deposition.

F. Solving a Growing Problem - Short -Term Solutions

    1. Many stopgap measures have been used to reduce the threat of acid deposition. These include the use of smokestack scrubbers, combustion of low-sulfur or desulfurized coal, and liming lakes to neutralize acidity.

G. Long-Term Sustainable Strategies

    1. Long-term sustainable strategies include: fuel efficiency, renewable fuels, recycling, population stabilization, and growth management.

III. Global Warming/Global Climate Change - Scientific evidence suggests that human activities can affect local climate.

A. Global Energy Balance and the Greenhouse Effect

    1. Ordinarily, the earth's energy input is offset by its energy output, thus, a balance is maintained.

B. Upsetting the Balance: Global Warming and Global Climate Change

    1. Various natural and anthropogenic gases trap heat in our atmosphere; the accumulation of these greenhouse gases (CO_2, methane, nitrous oxides, and CFCs) is thought to pose a threat of global warming.

    2. The concentrations of greenhouse gases have risen dramatically in the atmosphere in the past 45 years. The increased concentration of these gases is believed to be the cause of the increase in global temperature in the past 30 years.

C. Signs of Global Climate Change: Is the Planet Already Warming?

    1. Global warming is not science fiction; signs of global warming and climate change have already been observed.

D. Predicting the Effects of Greenhouse Gases

    1. Global climate models are used by scientists to predict possible climatic effects.

    2. Scientists predict a significant global increase in temperature by the end of the next century.

    3. Global temperature increases could shift rainfall patterns and have a profound effect on food production.

    4. Global warming might also be associated with an increase in the number and severity of storms.

E. The Ecological Effects of Global Climate Change

    1. Organisms could be profoundly effected by global climate change, particularly if the rate of change exceeds their ability for adaptation.

F. Uncertainties: What We Don't Know

    1. Current climate models are not able to predict with accuracy what specific effects global warming will have.

    2. It is possible that warming of the world's oceans, melting of land-based ice, and the loss of forests may result in a rapid increase in global carbon dioxide and accelerate the warming process. Other factors, however, may offset these changes.

G. Solving a Problem in a Climate of Uncertainty: Weighing Risks and Benefits

1. The cost of accepting the greenhouse hypothesis and acting accordingly is far lower than the potential cost of rejecting it.
   H. Solving the Problem Sustainably - Human systems need to be redesigned to foster sustainablility. The applying the principles of sustainability below will reduce the root causes of global warming:
      1. Population Stabilization and Restoration
      2. Recycling, Energy Efficiency and Renewable Energy
      3. International Cooperation to Halt Global Warming

Suggestions for Presenting the Chapter

- Instructors should emphasize the global impact of ozone depletion. Students should be aware of their impact concerning this problem.
- Instructors should examine how unsustainable human systems lead to acid deposition. Students should be encouraged to follow the principles of sustainability outlined in the text.
- The potential impacts of global warming should be stressed during lecture and class discussions. Students should be encouraged to locate the latest information about global climate change.
- Students might start monitoring your local air quality as a class project. There may already be a monitoring program in your area that you could participate in.
- Recommended web sites:

  National Climatic Data Center/NOAA: http://www.ncdc.noaa.gov/
  World Resources Institute/Climate Change: http://www.wri.org/climate/
  United Nations Environmental Program/Climate Change:
                  http://www.unep.org/Documents/Default.asp?ClassID=10000
  Global Warming Information Page: http://www.globalwarming.org/
  World Wildlife Federation's Climate Change Campaign: http://www.panda.org/climate/

True/False Questions

1. ___ Chloroflurocarbon molecules can destroy ozone molecules in the atmosphere.
2. ___ Ozone screens out 50% of the sun's harmful ultraviolet radiation.
3. ___ Ultraviolet radiation is potentially harmful to life on earth.
4. ___ Freon-12 is used in refrigerators and air conditioners as a refrigerant.
5. ___ The ozone layer in the atmosphere is also called the ionosphere.
6. ___ Carbon tetrachloride is an ozone-depleting solvent that was once widely used.
7. ___ The highest level of ozone depletion on Earth occurs over the North Pole.
8. ___ Excess ultraviolet radiation can cause skin burns, cataracts and skin cancer.
9. ___ Phytoplankton are not effected by increased ultraviolet radiation levels.
10. ___ It will take 100-200 years for the ozone layer to fully recover from current damage.
11. ___ An acid is a substance that adds hydrogen ions to a solution.
12. ___ Acidity is measured on the Rockwell scale.

13.___ Wet deposition refers to the evaporation of water from lakes and rivers.
14.___ Acidity in the soil can cause the leaching of heavy metals from the soil into surface water.
15.___ About 70% of all anthropogenic sulfur dioxide comes from electric power plants.
16.___ Baking soda and lime are very acidic substances.
17.___ Rain with a pH of 4 is 10 times more acidic than rain with a pH of 5.
18.___ The ozone layer is a nonrenewable form of protection that converts harmful ultraviolet radiation into heat.
19.___ Most CFC s released in the atmosphere come from natural sources like volcanoes and evaporation.
20.___ Acid deposition from pollutants is a local problem causing very limited and minimal social, economic and environmental impacts.

True/False Key:

1. T 2. F 3. T 4. T 5. F 6. T 7. F 8. T 9. F 10. T 11. T 12. F 13. F 14. T 15. T 16. F 17. T 18. F 19. F 20. F

Fill-in-the-Blank Questions

1. The pH scale ranges from 0 to ___.
2. The pH value indicating a neutral solution is ___.
3. ___ deposition refers to acids deposited in rain and snow.
4. ___ deposition occurs when air pollutants settle out of the atmosphere.
5. _____ are chemical substances that allow aquatic systems to resist changes in pH.
6. _____ interferes with normal calcium deposition in bird eggs resulting in soft eggshells.
7. Acid deposition may damage the foliage and ____ of plants.
8. Approximately ___-___ of the sunlight striking the Earth and its atmosphere is reflected back into space.
9. Carbon dioxide molecules in the atmosphere absorb _____ radiation escaping from the Earth's surface and radiate it back to the Earth.
10. The trapping of heat within the Earth's atmosphere by pollutants is called the _____ effect.
11. The four most important gases involved in global warming are: carbon dioxide, chlorofluorocarbons, nitrous oxide and _____.
12. One molecule of CFC is equivalent to _____ molecules of carbon dioxide.
13. In 1987, 24 nations signed a treaty called the _____ Protocol which would cut production of five CFCs by half by 1999.
14. The ozone layer extends from 10 to __ miles above the Earth's surface.
15. The ozone layer is threatened by the use of CFCs and ___ travel through the atmosphere.
16. All jets release ____ ____ gas that can react with ozone in the upper atmosphere.
17. In small amounts, _____ radiation tans the skin and stimulated vitamin D production.

18. The greatest declines in ozone have been recorded over _____ and the southern tip of Argentina.

19. CFCs are being replaced by a class of compounds called _____ that are much less damaging to atmospheric ozone.

20. The most widely used CFC is Freon-___.

Fill-in Key:

1. 14
2. 7
3. wet
4. dry
5. buffers
6. aluminum
7. roots
8. one-third
9. infrared
10. greenhouse
11. methane
12. 15,000
13. Montreal
14. 30
15. jet
16. nitric oxide
17. ultraviolet
18. Antarctica
19. HCFCs
20. 12

Multiple Choice Questions

1. This layer of gas protects life on earth from the sun's harmful ultraviolet radiation:
    a. oxygen layer
    b. carbon dioxide layer
    * c. ozone layer
    d. CFC layer
    e. nitrogen layer

2. This layer of the atmosphere contains gas which absorbs harmful ultraviolet light:
    a. lithosphere
    b. ionosphere
    * c. stratosphere
    d. troposphere
    e. hydrosphere

3. This group of chemicals break down in the upper atmosphere and destroy the gas that absorbs ultraviolet light; they are popular refrigerants:
    a. ozone
    * b. chlorofluorocarbons
    c. ammonia
    d. alcohols
    e. noble gases

4. Which of the following activities deplete the ultraviolet absorbing gas in the upper atmosphere?
   a. chlorofluorocarbon release into the air
   b. high flying aircraft release nitric oxide
   c. release of bromine into the air
   d. a and b only
 * e. a, b, and c

5. The greatest decline in the gas layer that absorbs ultraviolet radiation in the upper atmosphere occurs in winter months over:
   a. New Mexico
 * b. Antarctica
   c. New Zealand
   d. Nova Scotia
   e. Germany

6. Excessive exposures to ultraviolet radiation in humans can result in:
   a. cataracts
   b. skin cancer
   c. immune system suppression
   d. sun burns
 * e. all of the above

7. Chlorofluorocarbons take __ years to migrate to the upper atmosphere where they have their harmful effects.
   a. 5
   b. 10
 * c. 15
   d. 20
   e. 25

8. At least ___ years will be required to return ozone levels in the upper atmosphere back to 1985 levels.
   a. 25
   b. 50
   c. 75
 * d. 100
   e. 200

9. An acid is a chemical that adds _____ ions to a solution.
   a. sodium
   b. potassium
   c. chloride
 * d. hydrogen
   e. hydroxyl

10. Substances with a pH less than 7 are:
 * a. acidic.
   b. basic.
   c. neutral.
   d. like water.
   e. nonreactive.

11. A solution with a pH of 7 contains ___ times more hydrogen ions than a solution of pH 10.
   a. 10
   b. 100
 * c. 1000
   d. 10,000
   e. 100,000

12. A solution with a pH of 7 contains ___ times fewer hydrogen ions than a solution of pH 6.
 * a. 10
   b. 100
   c. 1000
   d. 10,000
   e. 100,000

13. The pH scale runs from :
   a. 0-7
   b. 0-10
 * c. 0-14
   d. 1-10
   e. 1-14

14. The pH of "pure " rain would be approximately:
   a. 3.5
 * b. 5.7
   c. 7.0
   d. 8.5
   e. 9.0

15. The pH of rain in unpolluted areas is slightly acidic because it has some ____ dissolved in it.
   a. oxygen
   b. nitrogen
   c. water vapor
 * d. carbon dioxide
   e. sulfuric acid

16. Wet deposition refers to acids deposited by:
    a. particulates falling to earth.
    * b. rain and snow
    c. ocean spray
    d. automobile exhaust
    e. animal metabolism

17. Dry deposition refers to acids deposited by:
    * a. particulates falling to earth.
    b. rain and snow.
    c. ocean spray.
    d. automobile exhaust.
    e. animal metabolism.

18. Sulfur oxide and sulfates can combine with water to form:
    a. carbonic acid
    * b. sulfuric acid
    c. nitric acid
    d. hydrochloric acid
    e. muriatic acid

19. Natural sources of sulfur dioxide include:
    a. forest fires
    b. volcanoes
    c. power plants
    * d. a and b only
    e. a, b, and c

20. About ___% of all sulfur dioxide comes from electric power plants that burn coal.
    a. 15
    b. 35
    c. 50
    * d. 70
    e. 90

21. Chemical substances that resist change in pH are called:
    a. acids
    b. bases
    c. alkali
    * d. buffers
    e. salts

22. Acid deposition affects aquatic systems by:
    a. killing aquatic organisms.
    b. impairing reproduction.
    c. impairing growth.
    d. increasing the concentration of metals in the water.
    * e. all of the above

23. Acid deposition is damaging to trees because it:
    a. impairs germination of some species.
    b. damages the leaves.
    c. may affect the root systems of trees.
    d. a and b only
    * e. a, b, and c

24. Acid deposition is damaging to crops because it:
    a. damages the leaves.
    b. may impair growth.
    c. may impair photosynthesis
    d. may alter the soil
    * e. all of the above

25. The 1990 amendments to the Clean Air Act require that by the year 2010 sulfur dioxide emissions are to be ___% below 1980 levels.
    a. 20
    * b. 40
    c. 60
    d. 80
    e. 95

26. International efforts to control sulfur dioxide emissions rely on which of the following strategies?
    a. Installation of scrubbers on new and existing coal-fired power plants.
    b. Combustion of low-sulfur coal or natural gas in utilities.
    c. Combustion of desulfurized coal.
    d. a and b only
    * e. a, b, and c

27. Which of the following strategies will reduce acid deposition from air pollution?
    a. Using fuel efficiently.
    b. Using renewable fuels.
    c. Stabilizing population growth.
    d. Growth management.
    * e. all of the above

28. Much of the sunlight reaching the earth and its atmosphere is converted into heat and is eventually radiated:
* a. back into space.
   b. back to the earth.
   c. into the air.
   d. and absorbed by ocean water,
   e. and absorbed by living organisms.

29. Heat is called _____ radiation.
   a. ultraviolet
   b. ionizing
   c. nuclear
* d. infrared
   e. gamma

30. Chemical substances that increase the Earth's surface temperature are called:
   a. heat sinks
   b. fuels
* c. greenhouse gases
   d. anthropogenic factors
   e. particulates

31. Which of the following substances is involved in warming the earth's atmosphere?
   a. water vapor
   b. carbon dioxide
   c. nitrous oxide
   d. methane
* e. all of the above

32. Which of the following substances are **not** involved in warming the earth's atmosphere?
   a. methane
   b. chlorofluorocarbons
   c. carbon dioxide
   d. nitrous oxide
* e. helium

33. The trapping of heat within the Earth's atmosphere by various pollutants is called the:
   a. photoelectric effect.
   b. Coriolis effect.
   c. Gaia effect.
* d. greenhouse effect.
   e. Doppler effect.

34. The largest single contributor to global warming in the atmosphere is:
   a. chlorofluorocarbons
   b. methane
   c. nitrous oxide
* d. carbon dioxide
   e. ozone

35. Which of the following is a sources of methane in the atmosphere?
   a. wetlands
   b. rice fields
   c. fossil fuels
   d. livestock
* e. all of the above

36. Scientists from the United Nations recently predicted a __ degree Celsius increase in the average global temperature from 1990 to 2100.
   a. 0.5
   b. 1
* c. 2
   d. 3
   e. 4

37. Which of the following responses is a possible effect of global warming?
   a. Raising sea levels.
   b. Rainfall patterns could change.
   c. River flows and groundwater could decrease in some areas.
   d. The number and severity of storms could increase.
* e. all of the above

38. Deforestation is responsible for about _____ of the annual global increase in carbon dioxide.
   a. one-eighth
* b. one-fourth
   c. one-third
   d. one-half
   e. two-thirds

39. Which of the following containers consumes the least amount of energy per use?
    a. aluminum can, used once
    b. recycled steel can
    c. glass bottle, used once
    d. recycled glass bottle
*   e. refillable glass bottle, used 10 times

40. Which of the following activities would be expected to reduce atmospheric carbon dioxide levels?
    a. Constructing new coal burning power plants.
*   b. Replanting large sections of tropical rainforest.
    c. Harvesting more timber for wood and paper production.
    d. Decreasing the fuel efficiency of automobiles.
    e. Allowing the world population to continue to increase at the current rate.

22

Water Pollution: Sustainably Managing a Renewable Resource

Chapter Outline

The Pollution of Surface Waters
   Sources of Water Pollution
   Point and Nonpoint Sources
   Organic and Inorganic Nutrients
   Infectious Agents
   Toxic Organic Water Pollutants
   Toxic Inorganic Water Pollutants
   Sediment
   Thermal Pollution

Groundwater Pollution
   Effects of Groundwater Pollution
   Cleaning Up Groundwater

Ocean Pollution
   Oil in the Seas
   Plastic Pollution
   Medical Wastes and Sewage Sludge
   Red Tide
   The Case of the Dying Seals

Water Pollution Control
   Legislative Controls
   Controlling Nonpoint Pollution
   Preventing Groundwater Pollution
   Water Pollution Control Technologies: End-of-Pipe Approaches
   Sustainable Solutions

Key Terms

water pollution

cross-media contamination

biochemical oxygen demand

natural succession

sedimentation

groundwater pollution

end-of-pipe controls

tertiary treatment

point source

organic nutrients

cultural eutrophication

infectious agents

streambed aggradation

ocean pollution

primary treatment

pretreatment

nonpoint source

inorganic nutrients

natural eutrophication

chlorinated organics

thermal shock

red tide

secondary treatment

Objectives

1. List the major types, sources and environmental effects of water pollution.
2. Define the following terms and give an example of each: "point source", "nonpoint source", "organic nutrient", and "inorganic nutrient".
3. Discuss natural and cultural eutrophication of lakes.
4. List the important infectious agents that are found in polluted water.
5. Discuss the major toxic chemical pollutants in surface and groundwater.
6. Summarize the major sources of ocean pollution and their environmental effects.
7. Summarize the major provisions of the Clean Water Act.
8. Discuss the current water pollution control technologies (end-of-pipe approaches).
9. List some sustainable solutions to pollution of surface and groundwater.

Lecture Outline

I. The Pollution of Surface Waters - Water pollution is any physical or chemical change in water which adversely affects organisms.
    A. Sources of Water Pollution
        1. Water pollutants arise from natural and anthropogenic causes. Pollutants can travel freely from one location to another through rivers, streams, and groundwater.
    B. Point and Nonpoint Sources
        1. Point sources are discrete, easily identified and controlled locations of concentrated pollution discharge.
        2. Nonpoint sources are large, less discrete areas over which dispersed pollutants are generated and discharged.
        3. The major nonpoint pollution source in the U.S. today is agriculture.
    C. Organic and Inorganic Nutrients- Nutrients in excessive amounts become pollutants.
        1. Organic Nutrients
            a. These include feedlot and slaughterhouse wastes, sewage treatment effluent, and some industrial wastes.
            b. Organic nutrients are oxygen demanding substances since their metabolism by bacteria is aerobic.

      c. The greater the organic nutrient load of water, the greater the biochemical oxygen demand.

    2. Inorganic Nutrients - Nitrates and Phosphates

      a. These nutrients, primarily nitrogen and phosphorus, stimulate the growth of aquatic plants; this can result in excess aquatic plant growth and oxygen depletion.

      b. Nitrogen fertilizer and phosphate-containing laundry detergents are major sources of inorganic nutrient pollution in freshwater systems.

    3. Eutrophication and Natural Succession

      a. Eutrophication is nutrient enrichment of a body of water, it can be either natural or cultural.

      b. Eutrophication and erosion result in the natural succession of lakes and ponds into swamps and, finally, dry land; humans have greatly accelerated this process.

D. Infectious Agents

    1. Pathogenic organisms may enter water through sewage effluent, animal wastes and processing byproducts, and certain wildlife species.

E. Toxic Organic Water Pollutants

    1. Thousands of these may contaminate water and pose aesthetic, environmental, or human health problems.

F. Toxic Inorganic Water Pollutants

    1. Mercury

      a. A common toxic heavy metal, mercury is widespread and harms aquatic organisms and humans.

    2. Nitrates and Nitrites

      a. These come from fertilizer and animal and human wastes; in high concentrations, they can poison infants who drink contaminated water.

    3. Salts

      a. Salts, often used on roads during winter, can kill nearby organisms and pollute ground and surface water.

    4. Chlorine

      a. A widely used disinfectant, chlorine can react with organic compounds to produce various carcinogens and mutagens in water.

G. Sediment

    1. The most voluminous water pollutant in the U.S., sediment is generated by forestry, agriculture, mining, and construction.

    2. Sedimentation aggravates pollution problems, fills in lakes and streams, , and damages human property.

H. Thermal Pollution

    1. Unnatural changes in water temperature which adversely affect organisms are instances of thermal pollution.

    2. Thermal pollution increases metabolism and decreases dissolved oxygen in aquatic systems and may interfere with reproduction and migration of some species.

    3. Power plants are the major sources of thermal pollution.

    4. A sudden, dramatic change in water temperature is thermal shock; this can be fatal for many organisms.

II. Groundwater Pollution
   A. Effects of Groundwater Pollution
      1. Groundwater is subject to contamination from a variety of surface sites and activities.
      2. Common pollutants include chlorides nitrates, heavy metals, hydrocarbons, pesticides, and organic solvents; many are known carcinogens.
   B. Cleaning Up Groundwater
      1. Since groundwater renewal and recovery is quite slow, prevention is the best method of protection.

III. Ocean Pollution
   A. Oil in the Seas -Sources include natural seepage, well blowouts, tanker and pipeline discharges, tanker spills and urban runoff
      1. Biological Impacts of Oil
         a. The effects of oil pollution vary with the amount and rate of release, location, and water temperature.
      2. Reducing the Number of Spills
         a. New tanker designs and operational procedures are aimed at minimizing the risk of major oil spills.
   B. Plastic Pollution
      1. Discarded plastics ensnare, entangle, starve, or suffocate hundreds of thousands of fish, marine mammals, and birds yearly.
   C. Medical Wastes and Sewage Sludge
      1. Illegal dumping of medical wastes and legal and illegal dumping of sewage and sewage sludge have caused environmental damage and pose a serious human health threat. Legislation aimed at stemming these practices has been enacted.
   D. Red Tide
      1. Outbreaks of microscopic and often highly toxic algae appear to be on the rise worldwide. These outbreaks may be caused by an increase in inorganic nutrient pollution from agriculture, industry, and the human population.
   E. The Case of the Dying Seals
      1. Massive seal die-off in the late 1980's caused by a virus may have resulted from immune system suppression caused by a common pollutant, PCBs.

IV. Water Pollution Control
   A. Legislative controls
      1. The Clean Water Act primarily regulates point source pollution
      2. Though enhanced sewage treatment can effectively reduce point source pollution, these gains are often offset by increasing nonpoint source pollution.
   B. Controlling Nonpoint Pollution
      1. Zoning ordinances, soil conservation programs and other legal requirements can help reduce nonpoint source pollution.
   C. Preventing Groundwater Pollution
      1. Many states, under EPA advisement, have developed programs to protect groundwater supplies through zoning and discharge regulation.
   D. Water Pollution Control Technologies : End-of-Pipe Approaches
      1. Primary Treatment

        a. Primary sewage treatment consists of mechanical screening and settling and removes large objects and solids.

    2. Secondary Treatment

        a. Secondary sewage treatment utilizes microorganisms to digest biodegradable organic matter.

    3. Tertiary Treatment

        a. Specialized filters and processes can remove chemicals left by secondary treatment; however, such tertiary sewage treatment is costly.

        b. Use of holding ponds, land disposal, and other innovative techniques can further reduce pollution from sewage treatment systems.

E. Sustainable Solutions

    1. Pollution prevention through conservation and recycling combined with new approaches for waste utilization and restoration to address existing pollution problems comprise the basics of a sustainable water pollution management system.

    2. Each of us can take steps to minimize our personal contribution to water pollution problems and to minimize our overall environmental impact.

Suggestions for Presenting the Chapter

- A visit to your local water treatment facility and sewage treatment facility are excellent ways to study how our public systems work.
- Instructors should emphasize the effectiveness and necessity of personal action in reducing our environmental impacts. The end of the chapter contains some excellent suggestions in Table 22-6 which can be discussed in class.
- The importance of nonpoint sources of pollution should be discussed. The potential impacts of agriculture on water quality are particularly important. The relationship between farming practices/soil conservation and water quality should be stressed.
- An examination of local groundwater contamination is always interesting. Your state agencies can provide useful information about the water in local aquifers.
- Recommended web sites:

U.S. EPA Office of Water : http://www.epa.gov/ow/

U.S. EPA Office of Water/Ground Water and Drinking Water:
        http://www.epa.gov/OGWDW/

U.S. EPA Nonpoint Source Pollution Homepage: http://www.epa.gov/OWOW/NPS/

Water Environment Federation: http://www.wef.org/publicinfo/ (click on the "Go with the flow" link)

U.S. EPA Water Quality Report State by State: http://www.epa.gov/OWOW/305b/

True/False Questions

1. ___ Point sources of pollution are discrete locations and are easy to identify.
2. ___ Silvaculture is the technique used to mine silver in underground deposits.
3. ___ Farms, forests, lawns and urban streets are common nonpoint sources of pollution.

4. ___ The movement of a pollutant from air to water is known as cross-media contamination.
5. ___ The organic nutrient concentration in water can be estimated by measuring the pH of the water.
6. ___ Laundry detergent from households is the major anthropogenic source of plant nutrients in fresh waters.
7. ___ Inorganic fertilizers from cropland are the second most important anthropogenic source of plant nutrients in fresh waters.
8. ___ The accumulation of nutrients in lakes is called eutrophication.
9. ___ Monitoring nitrates is the best way to monitor levels of fecal contamination in water.
10. ___ The deposition of sediment in surface waters is called chlorination.
11. ___ Sudden changes in water temperature can cause fish to undergo thermal shock.
12. ___ Sedimentation in streams may result in a gradual widening if the channel, a process known as streambed aggradation.
13. ___ Thermal pollution of water increases the dissolved oxygen content of the water.
14. ___ About 50% of the crude oil in a spill is volatile and evaporates within three months.
15. ___ The Oil Pollution Act established a $1 billion fund to be used to clean up oil spills and pay for damages.
16. ___ Millions of tons of plastic are dumped into the ocean each year, killing hundreds of thousands of marine mammals, fish and birds.
17. ___ The phocine distemper virus was responsible for deaths of black ducks off the coast of Denmark in the late 1980s.
18. ___ The Clean Water Act is the most important group of water laws in the United States.
19. ___ Primary sewage treatment removes large objects and settles out organic matter or sludge in a settling tank.
20. ___ Pretreatment is a technique to remove toxic substances, like metals, from water before sewage treatment.

True/False Key:

1. T 2. F 3. T 4. T 5. F 6. F 7. F 8. T 9. F 10. F 11. T 12. T 13. F 14. F 15. T 16. T 17. F 18. T 19. T 20. T

Fill-in-the-Blank Questions

1. The practice of using municipal sewage to fertilize crops is known as ____ disposal.
2. In contrast to fossil fuel power plants, photovoltaics and wind energy require little

   ____.
3. _____ treatment produces high quality water but is rarely used because it is costly.
4. _____ treatment uses bacteria and other organisms to decompose agitated waste.
5. _____ sources of pollution include city streets, lawns and farm fields.

6. ___ tide is a term referring to algal blooms in coastal waters that often are toxic to marine and human life.
7. About ___% of the crude oil in a spill is volatile and evaporates within three months.
8. ___ of the oil polluting the oceans comes from natural seepage; the rest comes from human sources.
9. Groundwater typically moves from 2 inches to __ feet a day.
10. _____ pollution can be controlled by constructing ponds for collecting and cooling water before its release into nearby lakes and streams.
11. Sudden changes in water temperature can cause fish to suffer _____ shock.
12. Streambed _____ results in a gradual widening of the channel.
13. The deposition of sediment in surface waters is called _____.
14. _____ is emitted from many sources and is one of the most common and most toxic inorganic pollutants.
15. The accumulation of nutrients in lakes is called _____.
16. Natural eutrophication and natural soil erosion can transform shallow lakes into swampland then into dry land, a process called natural _____.
17. ____% of stream pollution is coming from nonpoint sources.
18. BOD stands for _____ oxygen demand.
19. The second most important anthropogenic source of inorganic nutrient pollution in the U.S. is _____ detergents.
20. Lakes, streams and rivers are collectively referred to as _____ waters.

Fill-in Key:

1. land
2. water
3. tertiary
4. secondary
5. nonpoint
6. red
7. 25
8. half
9. 2
10. nonpoint
11. thermal
12. aggradation
13. sedimentation
14. mercury
15. eutrophication
16. succession
17. 65
18. biochemical
19. laundry
20. surface

Multiple Choice Questions

1. Which of the following is not classified as surface water?
   a. lakes
   b. streams
   c. rivers
   d. ocean
   * e. aquifers

2. Water pollution that arises from identifiable sources such as factories is called _____ pollution.
   * a. point source
   b. nonpoint source
   c. industrial
   d. chemical
   e. overt

3. The predominant source of nonpoint pollution is:
   a. mining
   b. construction
   c. silvaculture
   d. urban runoff
   * e. agriculture

4. The standard measurement of the organic nutrient concentration in a water sample is the:
   a. dissolved oxygen test
   b. Winkler test
   * c. biochemical oxygen demand
   d. Biuret test
   e. pH

5. The degradation of organic pollutants by bacteria under aerobic conditions uses up _____ in the water.
   a. carbon dioxide
   * b. oxygen
   c. iron
   d. methane
   e. carbon

6. Which of the following sources of water pollution is the largest in the United States?
   a. municipal
   b. industrial
   * c. nonpoint
   d. background
   e. other point sources

7. These substances are often the limiting factor for the growth of many plants:
   a. sulfur and nickel
   b. manganese and copper
   * c. nitrogen and phosphorus
   d. carbon and hydrogen
   e. water and chlorine

8. The most important source of inorganic nutrients for plants in freshwater is:
   a. laundry detergents
   * b. inorganic fertilizers used in agriculture
   c. atmospheric pollution
   d. organic decomposition
   e. landfills

9. The natural accumulation of nutrients in lakes is called:
   a. fermentation
   * b. natural eutrophication
   c. acidification
   d. aging
   e. cultural eutrophication

10. The natural process of a lake becoming a marsh and then dry land is called:
    a. cultural eutrophication
    b. restoration
    * c. succession
    d. evolution
    e. natural selection

11. Which of the following is a major source of infectious agents in water?
    a. Untreated or improperly treated sewage.
    b. Animal wastes in fields near waterways.
    c. Meat-packing and tanning plants that release untreated wastes into water.
    d. Some wildlife species that transmit waterborne diseases.
  * e. all of the above

12. Fecal contamination of water can be monitored by measuring the levels of _____ in the water sample.
    a. oxygen
    b. carbon dioxide
  * c. coliform bacteria
    d. amoeboe
    e. protozoans

13. About _____ of U.S. rivers now violate standards for coliform bacteria.
    a. one-eighth
    b. one-fourth
  * c. one-third
    d. one-half
    e. two-thirds

14. Which of the following responses is a legitimate concern about the impact of toxic chemicals in waterways?
    a. Many chemicals are nonbiodegradable and persist in the ecosystem.
    b. Some chemicals biomagnify in the food web.
    c. Some chemicals bioaccumulate.
    d. Some chemicals are carcinogenic.
  * e. all of the above

15. Metals, acids, and salts are _____ water pollutants.
    a. organic
  * b. inorganic
    c. ground
    d. innocuous
    e. nontoxic

16. Nitrate can be converted to the toxic ion nitrite in the:
  * a. human intestine.
    b. lungs.
    c. red blood cell.
    d. liver.
    e. skin

17. Chlorine is commonly used to:
    a. Kill bacteria in drinking water.
    b. Destroy harmful organisms in waste water treated by sewage treatment plants.
    c. Kill algae, bacteria, and fungi that grow inside and clog pipes of industrial systems.
    d. a and b only
  * e. a, b, and c

18. The leading water pollutant in the U.S. in terms of volume is:
    a. fertilizer
    b. phosphates
  * c. sediment
    d. sewage
    e. urban runoff

19. The filling of streambeds with sediment is called:
    a. subsidence
    b. succession
  * c. aggradation
    d. loading
    e. deposition

20. The U.S. electric power industry uses about ____% of all cooling water in the country.
    a. 26
    b. 46
    c. 56
    d. 76
  * e. 86

21. High temperature water released into waterways by industry is called_____ pollution.
    a. toxic
    b. instream
  * c. thermal
    d. slurry
    e. irrigation

22. Groundwater typically moves about 2 inches to _____ feet per day.
 * a. 2
   b. 20
   c. 200
   d. 2,000
   e. 20,000

23. About ____ metric tons of oil enters the world's seas every year.
   a. 1
   b. 2
 * c. 3
   d. 4
   e. 5

24. About half the oil that contaminates the sea comes from:
 * a. natural seepage from offshore deposits.
   b. well blowouts.
   c. breaks in pipelines.
   d. tanker spills
   e. inland oil disposal

25. About ___% of the oil spilled from human sources evaporates within 3 months.
 * a. 25
   b. 50
   c. 75
   d. 80
   e. 90

26. In 1990 the U.S. Congress passed the _____ which establishes a fund to be used to clean up oil spills and pay for damages.
   a. Resource Conservation and Recovery Act
 * b. Oil Pollution Act
   c. Clean Water Act]
   d. Occupational Safety and Health Act
   e. Toxic Substances Control Act

27. The tragic oil spill in Prince Edward Sound in Alaska happened in:
   a. 1970
   b. 1979
 * c. 1989
   d. 1990
   e. 1992

28. By the year _____ all single hull oil tankers will be banned from use.
   a. 1999
   b. 2005
 * c. 2015
   d. 2020
   e. 2025

29. The oil company involved in the spill in Prince William Sound is:
   a. Amoco
   b. Shell
   c. Texaco
 * d. Exxon
   e. Conoco

30. In 1988 the U.S. Congress passed the _____ which makes it unlawful for any U.S. vessel to discard plastic garbage into the seas.
   a. Ocean Dumping Ban Act
   b. Medical Waste Tracking Act
 * c. Plastic Pollution Control Act
   d. Oil Pollution Act
   e. Resource Conservation and Recovery Act

31. This law, passed by the U.S. Congress in 1988, prohibits the dumping of sewage sludge into the ocean:
 * a. Ocean Dumping Ban Act
   b. Medical Waste Tracking Act
   c. Plastic Pollution Control Act
   d. Oil Pollution Act
   e. Resource Conservation and Recovery Act

32. This law was a two year program designed to stop illegal dumping of medical wastes:
 * a. Ocean Dumping Ban Act
   b. Medical Waste Tracking Act
   c. Plastic Pollution Control Act
   d. Oil Pollution Act
   e. Resource Conservation and Recovery Act

33. The *red tide* refers to reddish _____ in coastal waters.
    a. fish released
    * b. algal blooms
    c. sewage released
    d. kelp beds
    e. bacteria

34. In 1988 the massive dieoff of harbor seals was caused by a virus but probably had a greater effect because:
    a. the seals were suffering from a bacterial infection at the same time.
    * b. the seals immune system was compromised by PCB pollution in the water.
    c. the North and Baltic seas have been clean for years.
    d. seals have weak immune systems.
    e. seals eat fish contaminated with the virus.

35. The Clean Water Act focuses primarily on:
    a. nonpoint source pollution
    b. point source pollution
    c. sewage treatment plants and factories
    d. a and b only
    * e. b and c only

36. The Clean Water Act provides funding for:
    a. the clean up of offshore oil drilling pollution.
    * b. construction and improvement of sewage treatment plants.
    c. preventing medical waste dumping in the oceans.
    d. controls the way oil is hauled in tankers.
    e. reduction in agricultural runoff.

37. The 1980 Helsinki Convention was the first international agreement to reduce marine pollution of the :
    a. Arctic Ocean
    b. Pacific Ocean
    c. Mediterranean Ocean
    d. Indian Ocean
    * e. Baltic Sea

38. The first sewage treatment plant in the United States was built in Memphis, Tennessee in:
    * a. 1880
    b. 1900
    c. 1910
    d. 1925
    e. 1945

39. This stage of sewage treatment removes much of the nitrogen and phosphorus by employing bacteria and other decomposers:
    a. primary treatment
    * b. secondary treatment
    c. tertiary treatment
    d. composting stage
    e. sludge stage

23
Pests and Pesticides: Growing Crops Sustainably

Chapter Outline

Chemical Pesticides
Modern Chemical Pesticides
Growth in the Use of Chemical Pesticides
Overuse
Biological Impacts of Pesticides
The Economic Costs of Pesticide Use
Herbicides in Peace and War
Peacetime Uses: Pros and Cons
The Alar Controversy: Apples, Alar, and Alarmists?

Controlling Pesticide Use
Bans on Pesticide Production and Use
Registering Pesticides
Establishing Tolerance Levels and Monitoring Produce

Integrated Pest Management: Protecting Crops Sustainably
Environmental Controls
Genetic Controls
Chemical Controls
Cultural Controls
Educating the World About Alternative Strategies
Governmental Actions to Encourage Sustainable Agriculture

Key Terms

chemical pesticides	first-generation pesticides	second-generation pesticides
broad-spectrum pesticides	narrow-spectrum pesticides	chlorinated hydrocarbons
organic phosphates	carbamates	genetically resistant insects
biomagnification	herbicides	insecticides
integrated weed management	agent orange	registration
licensed applicators	tolerance levels	integrated pest management
heteroculture	crop rotation	biological control
sterile male technique	third-generation pesticides	pheromones
confusion technique	juvenile hormone	molting hormone
cultural controls		

Objectives

1. Discuss the historical development of pesticides including all three "generations" of pesticides.
2. Discuss the three types of synthetic organic pesticides and give examples of each type.
3. Summarize the biological impact of chemical pesticide use.
4. Discuss the advantages and disadvantages of herbicide use.
5. Define the following terms: "integrated weed management" and "integrated pest management".
6. Discuss the effectiveness of the Federal Insecticide, Fungicide, and Rodenticide Act in regulating pesticide use.
7. Discuss the techniques used in integrated pest management.

Lecture Outline

I. Chemical Pesticides - Pest control measures have been used throughout the centuries. They have consisted of chemicals and cultural controls.
   A. Modern Chemical Pesticides
      1. Pesticides are pest-killing chemicals.
      2. First-generation pesticides were simple but mostly either toxic or ineffective.
      3. Second-generation pesticides began with DDT; these are synthetic organic compounds, of which thousands have been developed.
      4. Synthetic pesticides may be either broad- or narrow-spectrum, depending on their specificity.
      5. Pesticides fall into three chemical families: chlorinated hydrocarbons, organic phosphates, and carbamates.
   B. Growth in the Use of Chemical Pesticides
      1. Several million tons of pesticides are used each year, mostly in the developed nations. The bulk of the pesticides used are herbicide.
   C. Overuse
      1. Pesticides are often applied in excess which increases the danger of their use.
   D. Biological Impacts of Pesticides
      1. Harmful biological impacts of pesticides include: destruction of beneficial insects, development of genetically resistant pests, and health effects in nontarget organisms and humans, especially chemical and farm workers, rural residents, and consumers.
   E. The Economic Costs of Pesticide Use
      1. Pesticides have caused considerable economic damage especially from poisonings, death, and loss of wildlife.
   F. Herbicides in Peace and War
      1. Sixty percent of all pesticides used are herbicides.
   G. Peacetime Uses: Pros and Cons
      1. The benefits of herbicides are listed below.
         a. They decrease the amount of cultivation needed to control weeds and thereby reduce operating costs.

      b. They reduce weed damage when soils are too wet to cultivate because crops can be sprayed by plane

      c. They reduce water usage because water evaporates more quickly from ground that has been cultivated to control weed growth.

  2. The disadvantages of herbicides are summarized below.

      a. Resistant weeds may proliferate and necessitate further application of herbicide.

      b. Resistance to the herbicide is developed and creates a more severe problem.

      c. Herbicides encourage accumulation of material on the ground that is good for growth of pest populations (dead plant material). Herbicide use may actually increase insect pest problems necessitating the use of insecticides.

      d. Herbicides can reduce the farmer's incentive to rotate crops which can allow pests to increase.

      e. Herbicides can decrease some plants resistance to insects and disease by damaging the plant's tissues, changing their metabolic rates and effecting rate of growth. All of these factors could make plants more susceptible to insects and disease.

      f. Some herbicides are toxic and may cause birth defects, cancer, and other illnesses in animals including humans.

  3. Extensive use of chemical defoliants (Agent Orange) during the Vietnam war resulted in substantial environmental and health impacts. The human health effects are attributed to dioxins that contaminate the herbicide.

H. The Alar controversy: Apples, Alar, and Alarmists?

  1. Children may be at higher risk to pesticides than adults.

II. Controlling Pesticide Use

A. Bans on Pesticide Production and Use

  1. Bans on harmful pesticides in the United States have been effective. The ban on DDT has allowed the recovery of a number of endangered bird species.

  2. Bans on pesticides in the United States are only part of the answer. Many imported vegetables and fruits have been found with pesticide contamination.

  3. Global bans of harmful pesticides will be necessary to have significant impacts here and abroad.

B. Registering Pesticides

  1. The U.S. EPA registers newly developed and previously introduced pesticides for general or restricted use. They can stipulate what crops they can be used on in the registration process. Improvements are needed in the registration process. For instance, the process does not require testing for neurotoxicity or toxicity to the immune system.

C. Establishing Tolerance Levels and Monitoring Produce

  1. Tolerance levels are concentrations of pesticides in or on foods that are believed to be a tolerable health risk. The EPA sets these levels in the U.S..

  2. The U.S. Food and Drug Administration and state agricultural agencies are charged with enforcement of the EPA standards.

  3. The programs for monitoring/enforcement are under-funded and understaffed.

  4. The EPA ranks pesticides in food as one of the nation's most serious health concerns.

III. Integrated Pest Management: Protecting Crops Sustainably - This system calls for the integrated use of environmental , genetic, chemical, and cultural pest control; with properly educated farmers and increased pest monitoring, results can be impressive.

A. Environmental Controls - These alter the environment to disfavor the pests.
   1. Increasing Crop Diversity
      a. Heteroculture and crop rotation help prevent rapid growth of pest populations.
   2. Altering the Time of Planting
      a. This technique can thwart pests by removing their food supply.
   3. Altering Plant and Soil Nutrients
      a. By manipulating nutrient levels in soils and thus plants, some pests can be suppressed.
   4. Controlling Adjacent Crops and Weeds
      a. This can help control pests by either eliminating food or habitat for them or by luring them off more valuable crops.
   5. Introducing Predators, Parasites, and Disease Organisms
      a. Use of these techniques mimics or supplements natural biotic environmental resistance factors which regulate pest populations.

B. Genetic Controls
   1. Sterile Male Technique
      a. This involves releasing sterilized males of the pest species, which mate with wild females who thus do not produce offspring.
   2. Developing Resistant Crops and Animals
      a. Genetic engineering and artificial selection can lead to the development of pest resistant crops and livestock.

C. Chemical Controls
   1. Second-Generation Pesticides
      a. With judicious, timely, and appropriate application of low-toxicity, specific, and nonpersistent pesticides, many of the benefits of these chemicals can be retained without accompanying ecological and human health damage.
   2. Third-Generation Pesticides
      a. Pheromones, insect hormones, and natural insecticides can, in some cases, effectively control pests without unintended environmental or human damage.

D. Cultural Controls - These various techniques offer alternatives to harsh chemical pesticides and are often quite effective.
   1. Monitoring
      a. Frequent pest monitoring is a necessary prerequisite to successful IPM.

E. Educating the World About Alternative Strategies
   1. Both governments and nongovernmental organizations have an important role to play in educating growers and government officials about the alternatives to traditional pest management strategies.

F. Governmental Actions to Encourage Sustainable Agriculture
   1. Governments can help promote sustainable agriculture by providing low-cost crop insurance for farmers who are making the transition to integrated pest management.
   2. Governments can also help by developing organic certification programs. States like Colorado and California already have these programs working.

Suggestions for Presenting the Chapter

- Instructors should emphasize that there are alternatives to our use of conventional agricultural techniques. Integrated pest management as a part of an organic agricultural system is the wave of the sustainable future.
- Public and governmental support of sustainable agricultural practices is necessary to make the transition from our current unsustainable system to a sustainable agricultural system. Students need to be encouraged to support organic production.
- Recommended web sites:

  United Nations Food and Agriculture Organization: http://www.fao.org/
  U.S. EPA Office of Prevention, Pesticides and Toxic Substances:
          http://www.epa.gov/opptsfrs/home/welcome.htm
  Natural Resources Defense Council/Toxic Chemicals and Health:
          http://www.nrdc.org/health/default.asp
  The Extension Toxicology Network: http://ace.ace.orst.edu/info/extoxnet/
  Pesticide Action Network: http://www.igc.org/panna/

True/False Questions

1. ___ Pesticides are chemicals that kill insects, weeds, and other organisms.
2. ___ First-generation pesticides were simple preparations made of ashes, sulfur, arsenic and other toxic compounds.
3. ___ DDT is a state-of-the-art pesticide that is safe to use and does not persist in the environment.
4. ___ DDT is a second-generation pesticide.
5. ___ Parathion and malathion are chlorinated hydrocarbon pesticides.
6. ___ DDT is an organic phosphate pesticide.
7. ___ Organophoshates and carbamates have an advantage over the chlorinated pesticides since they are quickly degraded in the environment.
8. ___ Pyrethroids are a very toxic group of natural and synthetic insecticides.
9. ___ Most pesticides used are herbicides.
10. ___ Herbicides are pesticides used to kill plants.
11. ___ Pesticide use results in the formation of genetically resistant pest species.
12. ___ Although insecticide use had increased 10-fold since World War II, crop damage has doubled.
13. ___ Pesticides poison fish and other species outright and also biomagnify in the foodchain.
14. ___ One benefit of herbicides is that they decrease the amount of mechanical cultivation needed to control weeds.
15. ___ Dioxins are safe contaminants of the herbicides used in Agent Orange.
16. ___ Rachel Carson's, *Silent Spring*, supported the widespread use of pesticides.
17. ___ Pesticide registration is required by the Federal Insecticide, Fungicide and Rodenticide Act.
18. ___ Integrated pest management has a goal of reducing pest populations to levels that do not cause economic damage while protecting the environment.
19. ___ Low-value crops planted to attract pest are called "cash crops".

20.___ Biological control uses natural predators and parasites to control insect pest populations.

True/False Key:

1. T 2. T 3. F 4. T 5. F 6. F 7. T 8. F 9. T 10. T 11. T 12. T 13. T 14. T 15. F 16. F 17. T 18. T 19. F 20. T

Fill-in-the-Blank Questions

1. _____-_____ pesticides are natural compounds that are effective at controlling pests.
2. The control of crop pests using natural predators and parasites is known as _____ control.
3. Planting several crops side by side in fields is known as _____.
4. _____ control refers to a number of techniques that alter the biotic and abiotic conditions in crops, making them inhospitable to pests.
5. _____ are believed to be 100,000 times more potent than the tranquilizer thalidomide which caused many birth defects in Europe.
6. Agent _____ is a 50-50 mixture of 2,4-D and 2,4,5-T.
7. Most pesticides used today are _____, pesticides that kill plants.
8. Organophosphates and _____ are more rapidly degraded in the environment than chlorinated hydrocarbons.
9. DDT is a _____-generation pesticide.
10. About ___% of the pesticides applied in the U.S. are herbicides.
11. DDT is a _____-spectrum pesticide.
12. Insecticides often kill _____ insects which often result in a pest outbreak.
13. _____ workers are frequently exposed to the highest levels of pesticides.
14. _____ applicators consume much less herbicide than aerial spraying and create less environmental contamination.
15. _____ is the permitting procedure used by the EPA to determine on which crops the pesticide can be used.
16. _____ pest management is a set of alternative strategies to control pest populations by applying sustainable technologies.
17. _____ levels are concentrations of pesticides in or on foods that are believed to pose an acceptable heath risk.
18. Crop _____ is an agricultural technique that reduces soil erosion and controls pests.
19. Crops that are planted to attract pests away from more valuable crops are called _____ crops.
20. _____ are natural sex attractants released by insects.

Fill-in Key:

1. third-generation
2. biological
3. heteroculture
4. environmental
5. dioxins
6. Orange
7. herbicides
8. carbamates
9. second
10. 60
11. broad
12. predatory
13. farm
14. wick
15. Registration
16. Integrated
17. tolerance
18. rotation
19. trap
20. pheromones

Multiple Choice Questions

1. Chemical substances that kill insects, weeds, and a variety of other organisms that reduce crop production are called:
   a. insecticides
   b. herbicides
   * c. pesticides
   d. rodenticides
   e. fungicide

2. Killing insect pests by burning infested fields is called:
   a. first-generation pesticides
   b. chemical pesticides
   c. herbicide
   * d. cultural controls
   e. backfire

3. The first synthetic organic pesticide was:
   * a. DDT
   b. DDE
   c. dieldrin
   d. hydrogen cyanide
   e. chlordane

4. An example of an organophosphate pesticide would be:
   a. DDT
   b. dieldrin
   * c. malathion
   d. chlordane
   e. carbaryl

5. Ashes, sulfur, arsenic, and ground tobacco are:
   a. cultural controls
 * b. first-generation pesticides
   c. second-generation pesticides
   d. third-generation pesticides
   e. fourth-generation pesticides

6. DDT and dieldrin are:
   a. cultural controls
   b. first-generation pesticides
 * c. second-generation pesticides
   d. third-generation pesticides
   e. fourth-generation pesticides

7. About _____ of all pesticides used in the world are applied in the U.S.
   a. one-eighth
   b. one-sixth
 * c. one-fifth
   d. one-fourth
   e. one-third

8. Which of the following responses is a problem with application of pesticides?
   a. They often kill natural pest control agents like predatory insects.
   b. They result in genetically resistant pests.
   c. They kill pollinating insects.
   d. They poison fish and other wildlife.
 * e. all of the above

9. Some pesticides can become more concentrated as they move up the food chain, this phenomenon is called:
 * a. biomagnification
   b. bioaccumulation
   c. biofiltration
   d. biodegradation
   e. bioremediation

10. In the United States, at least _____ workers are seriously poisoned each year by pesticides.
    a. 15,000
    b. 30,000
 * c. 45,000
    d. 60,000
    e. 75,000

11. DDT would be classified as a:
    a. narrow-spectrum pesticide
 * b. chlorinated hydrocarbon pesticide
    c. organophosphate pesticide
    d. carbamate pesticide
    e. natural pesticide

12. The organophosphates and carbamates have one key advantage over the chlorinated hydrocarbon pesticides, they::
    a. are not very toxic.
 * b. break down more easily after application.
    c. are very cheap
    d. will kill plants.
    e. are not nervous system toxins.

13. The herbicides 2,4-D and 2,4,5-T are synthetic organic compounds similar in function to plant hormones called _____.
    a. dioxins
 * b. auxins
    c. estrogens
    d. growth hormone
    e. renin

14. The benefits of herbicides are:
    a. They decrease the amount of mechanical cultivation needed.
    b. They reduce weed damage.
    c. They help reduce water usage.
    d. a and b only
 * e. a, b, and c

15. A drawback of herbicides is:
    a. Some weeds are resistant to some herbicides necessitating additional spraying.
    b. Some weeds develop resistance.
    c. Reduced cultivation due to herbicide use may encourage proliferation of pest organisms in plants/debris on the ground.
    d. Herbicides may reduce a farmer's incentive to rotate crops.
 * e. all of the above

16. Agent orange is a 50-50 mixture of:
    a. DDT and DDE
    b. DDT and dieldrin
    c. 2,4-D and DDT
 * d. 2,4-D and 2,4,5-T
    e. 2,4-D and dioxin

17. ____% of all chemical pesticides applied to crops are herbicides.
    a. 20
    b. 40
  * c. 60
    d. 80
    e. 90

18. Health problems from Agent Orange exposure may be linked to a very potent toxin that was present as a contaminant called _____ .
    a. dieldrin
    b. DDT
  * c. dioxin
    d. strychnine
    e. mercury

19. This chemical was sprayed on apples to delay ripening so that the apples stay fresher. This chemical is no longer used because of its cancer risk:
    a. DDT
    b. DDE
    c. malathion
  * d. alar
    e. diazinon

20. The book, *Silent Spring*, is credited with raising awareness about the dangers of pesticide use; it was written by:
    a. Paul Ehlrich
    b. Stewart Udall
  * c. Rachel Carson
    d. Dr. Bruce Ames
    e. Carl Sagan

21. Bans on the production and use of pesticides in the U.S. is only part of the answer to our pesticide problem because:
    a. Pesticides are being imported into the U.S..
    b. We are using other legal chemicals instead.
    c. Pesticides are only one class of toxic chemical for industrial use.
    d. Industry is using more harmful chemicals covertly.
  * e. Many pesticides banned in the U.S. are still being used in heavy amounts in other countries around the world.

22. Agent Orange was sprayed on fields in:
    a. Orange County, California
    b. Florida
    c. orange groves
  * d. Vietnam
    e. Korea

23. Restricted-use pesticides can only be used by:
    a. industry
    b. private individuals
  * c. licensed applicators
    d. selected companies
    e. the EPA

24. Current pesticide registration procedures do not require companies to test pesticides for:
    a. environmental persistence
    b. neurotoxicity
    c. immune system effects
    d. a and b only
  * e. b and c only

25. The EPA sets acceptable levels of pesticide residues on food for human consumption; these are called:
    a. attainment levels
  * b. tolerance levels
    c. abatement levels
    d. application levels
    e. residue levels

26. The monitoring of pesticide residues on food in the U.S. is done by the:
    a. Department of Agriculture
  * b. U.S. Food and Drug Administration
    c. EPA
    d. Occupational Safety and Health Administration
    e. Department of Commerce

27. The Federal Insecticide, Fungicide, and Rodenticide Act provides for _____ of pesticides.
* a. registration
  b. federal testing
  c. import
  d. only restricted use
  e. phase out

28. The alternative to conventional management of pest problems is sustainable approach called:
  a. new age control
  b. fourth-generation pesticides
* c. integrated pest management
  d. crop rotation
  e. cultural controls

29. Planting several crops side by side in a field is called
  a. monoculture
* b. heteroculture
  c. crop rotation
  d. silvaculture
  e. horticulture

30. Heteroculture is a technique of integrated pest management that uses _____ control.
* a. environmental
  b. genetic
  c. chemical
  d. cultural
  e. organic

31. Alternating the crop planted in a particular field every year is called:
  a. cultivation
* b. crop rotation
  c. strip cropping
  d. row cropping
  e. heteroculture

32. Altering the time of planting is a technique of integrated pest management that uses _____ control.
* a. environmental
  b. genetic
  c. chemical
  d. cultural
  e. organic

33. The National Academy of Sciences issued a report in 1987 concluding that one in every ____ Americans will develop cancer as a result of pesticide contamination of food.
  a. 2.5
  b. 25
* c. 250
  d. 2,500
  e. 25,000

34. The introduction of large numbers of sterilized males into an insect pest population is an example of _____ control.
  a. environmental
* b. genetic
  c. chemical
  d. cultural
  e. organic

35. An example of a third-generation pesticide is:
  a. DDT
  b. mercury
  c. dieldrin
  d. pyrethrin
* e. pheromones

36. Insect hormones applied to crops offer which of the following advantages?
  a. They are biodegradable.
  b. They are nontoxic.
  c. There is low persistence in the environment.
  d. a and b only
* e. a, b, and c

37. Cultivating to control weeds in a field is a/an _____ control method.
  a. environmental
  b. genetic
  c. chemical
* d. cultural
  e. rotation

38. The typical insect life cycle is: egg, _____,
    pupa, and adult.
    a. fertilized egg
*   b. larvae
    c. caterpillar
    d. subadult
    e. Imago

39. Destroying insect breeding grounds is a/an
    _____ control method.
    a. environmental
    b. genetic
    c. chemical
*   d. cultural
    e. Manual

40. Insect hormones can be sprayed at breeding
    time to disrupt normal reproductive activity;
    this is called:
    a. genetic engineering
    b. cultural control
*   c. confusion technique
    d. genetic control
    e. environmental control

24

Hazardous and Solid Wastes: Sustainable Solutions

Chapter Outline

Hazardous Wastes: Coming to Terms with the Problem
    Love Canal: The Awakening
    The Dimensions of a Toxic Nightmare
    LUST - It's Not What You Think

Attacking Hazardous Wastes on Two Fronts
    The Superfund Act: Cleaning Up Past Mistakes
    What to Do with Today's Waste: Preventing Future Disasters
    Dealing with Today's Wastes: A Variety of Options
    Disposing of Radioactive Wastes
    Some Obstacles to Sustainable Hazardous Waste Management
    Individual Actions Count

Solid Wastes: Understanding the Problem

Solving a Growing Problem Sustainably
    The Traditional Approach
    Sustainable Options: The Input Approach
    The Throughput Approach: Reuse, Recycling, and Composting

Key Terms

hazardous wastes	Superfund	in-plant options
process manipulation	reuse	recycling
detoxification	land disposal	incineration
low-temperature decomposition	secured landfills	NIMBY syndrome
municipal solid wastes	output approach	input approach
throughput approach	dumps	landfills
ocean dumping	waste-to-energy plants	aseptic containers
price preference policies	compost	co-composting

Objectives

1. Discuss the national significance, implications, and effects of the *Love Canal* toxic waste problem.
2. List some of the environmental effects of improper waste disposal.
3. Define the acronym "LUST" and suggest how this environmental problem is being remediated.
4. Discuss the primary provisions and effectiveness of the Comprehensive Environmental Response, Compensation and Liability Act (CERCLA).
5. Discuss the primary provisions and effectiveness of the Resource Conservation and Recovery Act (RCRA).
6. Summarize current sustainable options for reducing our hazardous wastes.
7. List some personal actions that will reduce hazardous wastes.
8. Define the following terms: "municipal solid wastes", "dumps", "landfills".
9. Discuss the composition of solid waste in the United States.
10. Summarize the current strategies for dealing with solid wastes.
11. Define "reuse" and "recycling" and give examples of solid wastes that can be handled by these techniques.
12. List some obstacles to recycling in our society and how these obstacles might be overcome.
13. Define "composting" and list the advantages of composting as a solid waste reduction strategy.

Lecture Outline

I. Hazardous Wastes: Coming to Terms with the Problem
   A. Love Canal: The Awakening
      1. Hazardous wastes are byproducts of industry that pose a threat to the environment.
      2. Love Canal is a tragic public health and environmental disaster, it symbolizes our past disregard for the basic rule that "there is no away."
      3. As a result of negligent (though legal) dumping of toxic wastes at Love Canal, the health and investments of nearly a thousand families were damaged or destroyed and hundreds of millions of dollars have been spent for compensation and cleanup.
   B. The Dimensions of a Toxic Nightmare
      1. The huge amount of waste produced and past irresponsible disposal practices combine to create a toxic waste nightmare.
      2. In the past, toxic wastes were disposed of largely by use of two criteria: cost and convenience; effects of this negligence include groundwater and soil contamination, human and other animal disease and death, costly cleanup attempts and remediation efforts, and forced abandonment of homes and small towns.
   C. LUST- It's Not What You Think
      1. Leaking underground storage tanks, "LUST," pose serious threats to groundwater worldwide.
      2. Costs of LUST correction may be prohibitively high.

II. Attacking Hazardous Wastes on Two Fronts
  A. The Superfund Act: Cleaning Up Past Mistakes
    1. The Resource Conservation and Recovery Act (RCRA) or "Superfund" is legislation aimed at improving waste-disposal practices in the U.S.
    2. Though well-intentioned, RCRA has its drawbacks, including high costs and enforcement difficulties and stimulation of illegal dumping.
    3. Alternative funding for cleanup and proper disposal through a tax or generators fee might allow the EPA to more effectively direct cleanup efforts.
  B. What to Do With Today's Waste: Preventing Future Disasters
    1. Preventing Leaking Underground Storage Tanks
      a. RCRA amendments may effectively address this problem.
    2. Weaknesses in RCRA
      a. Difficulties in administering and enforcing federal law, combined with underfunding and increasing responsibilities, have made it difficult for the EPA to close the loopholes in this law.
    3. Exporting Toxic Troubles
      a. Exporting toxic wastes does not address the root causes of our toxic waste problems and has serious ethical implications.
  C. Dealing With Today's Wastes: A Variety of Options
    1. Process Manipulation, Reuse, and Recycling
      a. These approaches help reduce waste output as well as reduce resource consumption.
    2. Conversion to Less Hazardous or Nonhazardous Substances
      a. Various technologies for detoxification, incineration and low-temperature decomposition may be used to help detoxify or destroy hazardous wastes.
    3. Perpetual Storage
      a. While appealing to those seeking a quick solution to the hazardous waste problem, resort to perpetual storage techniques are seen by many as a temporary and socially unjust solutions.
    4. Barriers to Waste Reduction
      a. Powerful incentives which encourage waste production remain in place today.
  D. Disposing of Radioactive Wastes
    1. Disposing of High-Level Radioactive Wastes
      a. Currently the U.S. has no disposal facility for high level radioactive wastes generated by weapons manufacture and nuclear energy production.
      b. By law, the Federal government must choose a waste disposal site, build and license a repository there, and begin accepting high-level wastes within a decade; public opposition, though, may delay this.
    2. Disposing of Low- and Medium-Level Radioactive Wastes
      a. Low-level radioactive wastes are currently landfilled in Nevada, Washington, and South Carolina.
      b. There is no repository yet for medium-level radioactive waste, though DOE has built a Waste Isolation Pilot Plant in New Mexico which might be licensed to receive these wastes.
  E. Some Obstacles to Sustainable Hazardous Waste Management
    1. Overall, while total amounts of toxic waste generated in the past few years have decreased, disposal problems have intensified.

2. The move away from land- and water-based disposal practices has created a new hazard, toxic waste incineration.
F. Individual Actions Count
   1. By reducing our participation in the production and improper disposal of hazardous wastes, we can minimize our personal contribution to this problem.

III. Solid Wastes: Understanding the Problem
   A. The huge amounts of solid waste we generate represent a squandering of energy, resources, and money; it is a result of population growth, increasing consumerism and affluence, and a combination of marketing, governmental, and societal factors which stimulate consumption and waste generation.
   B. The goals of modern solid waste management focus on source reduction and minimum-impact disposal.

IV. Solving a Growing Problem Sustainably
   A. The Traditional Strategy: Output Approach - This strategy focuses on ways to safely and economically dispose of waste.
      1. Dumps and
         a. While sanitary landfills are environmentally preferable to open dumps, they are still energy- and resource-intensive, potentially polluting, and generally undesirable as neighbors.
      2. Ocean Dumping
         a. This practice raises serious concerns over ecosystem damage, toxin biomagnification, and aesthetic degradation of oceans and beaches.
      3. Incineration
         a. Waste-to-energy plants, or incinerators, can reduce waste volume and recover energy from wastes, but they emit toxic air pollution and generate hazardous ash for disposal.
   B. Sustainable Options: The Input Approach
      1. Increasing Product Durability
         a. Making goods more durable will reduce resource use and waste generation.
      2. Reducing the Amount of Material in Products and Packaging
         a. Redesigning products to reduce waste is practical and economical.
      3. Reducing Consumption
         a. By rejecting the accelerating consumerism of our culture, individuals can reduce the amount of waste they generate.
   C. The Throughput Approach: Reuse, Recycling, and Composting - This largely consists of reusing and recycling materials before they enter the waste stream.
      1. The Reuse Option
         a. Many items commonly thrown out as waste can be reused one or more times. By reducing waste, reuse lessens all the problems associated with waste disposal.
      2. The Recycling Option
         a. Recycling a product to make more products saves energy, resources money, pollution, jobs, and aesthetics.
      3. Obstacles to Recycling
         a. There are few technological obstacles to recycling; rather, the barriers are political governmental, economic, and attitudinal.
      4. Overcoming the Obstacles

       a.   Spiraling landfill costs, energy and resource shortages, and pollution problems will encourage recycling efforts in the U.S..

   5. Procuring Recycled Materials
       a.   High recycling rates can be achieved only if a strong market for recycled products is maintained.

Suggestions for Presenting the Chapter

- Instructors should emphasize the virtues of conservation and recycling.
- A visit to a local landfill site is an excellent exercise to illuminate the concepts in this chapter. The cost of citing landfills and the associated environmental problems can be discussed as a part of this activity.
- A discussion of any local Superfund sites can be explored in class or as part of a fieldtrip.
- Visiting a local recycling facility is an excellent fieldtrip for your class. Students can see that recycling can have a real impact on the quality of the environment and support a sustainable economy.
- Recommended web sites:

   Love Canal USA: http://web.globalserve.net/~spinc/atomcc/lovecana.htm
   Chernoybl - The Accident and Progress Since 1986: http://www.uilondon.org/chernidx.htm
   Office of Civilian Radioactive Waste Management/DOE:
            http://www.rw.doe.gov/homejava/homejava.htm
   U.S. EPA Office of Solid Waste/RCRA and CERCLA Information:
            http://www.epa.gov/epaoswer/hotline/
   United Nations Environmental Program/Basel Convention on Hazardous Wastes:
            http://www.unep.ch/basel/

True/False Questions

1. ___ Combining sewage sludge with municipal solid waste for natural decomposition is called co-composting.
2. ___ A hidden cost that is being paid by the public from pollution or unsustainable practices is called an economic externality.
3. ___ Paper recycling uses one-third to one-half as much energy as the conventional process of making paper from wood pulp.
4. ___ It is cheaper to make a new aluminum can from bauxite ore than from recycled aluminum.
5. ___ Decreasing the life span of a product is a good source reduction technique to reduce solid waste.
6. ___ Reducing the amount of materials in good and their packaging will reduce solid wastes.
7. ___ Waste-to-energy plants burn garbage as a method of reducing municipal waste volume.

8. ___ The Marine Protection, Research and Sanctuaries Act ended dumping of sewage sludge at sea.
9. ___ A sanitary landfill is really just an open garbage dump.
10. ___ The largest category of municipal solid waste by weight is paper and cardboard.
11. ___ Phytoremediation is the use of sunlight to degrade otherwise toxic materials.
12. ___ Low-temperature decomposition is a technique that can be used to degrade radioactive wastes.
13. ___ Secured landfills are excavated pits lined by impermeable synthetic liners.
14. ___ Landfills are one of the most expensive waste disposal options.
15. ___ Substitution involves the use of nontoxic substitutes in manufacturing to reduce hazardous wastes.
16. ___ Hazardous wastes are waste products from homes, factories, businesses and military installations that pose a threat to people and the environment.
17. ___ Much of the money allocated to the Superfund Act has gone to legal fees.
18. ___ The Resource Conservation and Recovery Act was designed to cleanup Superfund sites around the United States.
19. ___ Times Beach was a town in Missouri that was purchased as a Superfund site.
20. ___ Under the Superfund Act the Department of Energy is in charge of collecting fines from parties responsible for hazardous waste contamination.

True/False Key:

1. T 2. T 3. T 4. F 5. F 6. T 7. T 8. F 9. F 10. T 11. F 12. F 13. T 14. F 15. T 16. T 17. T 18. F 19. T 20. F

Fill-in-the-Blank Questions

1. CERCLA is the acronym for the law that is often called the _____ Act.
2. The ___ is the federal agency that is in charge of Superfund activities.
3. The Resource Conservation and Recovery Act established a nationwide reporting system to monitor _____ wastes from production to disposal.
4. _____ is the burning of hazardous wastes in high-temperature furnaces.
5. Cyanide and toxic organic chemicals can be treated using low-temperature _____.
6. _____ repositories are needed for the storage of radioactive wastes.
7. The use of plants to remove hazardous materials from soil and water is called _____.
8. In the United States ___ % of the municipal solid waste is buried in landfills.
9. Open garbage dumps have been replaced by _____ landfills.
10. _____ gas produced in landfills can be drawn off by pipes and used as fuel.
11. Incinerators are also called _____-to-_____ plants.
12. Increasing product life to reduce the amount of waste is a strategy used in _____ reduction.
13. _____ materials and products reduces resource demand and offers other environmental benefits.

14. _____ refers to the return of materials to manufacturers for production of new products.
15. _____ is the process in which nutrients from organic wastes are returned to the soil.
16. _____ is a nutrient rich material much like humus and can be used to build soil fertility.
17. Mixing sewage sludge with municipal solid wastes (leaves, paper, grass clippings) for decomposition is called __-_____.
18. A hidden cost due to environmental damage or pollution that is paid by the public is called an economic _____.
19. The Ocean Dumping Ban Act phased out ocean dumping of _____ sludge.
20. The Nuclear Waste Policy Act established a timetable for the selection of an underground disposal site for high-level _____ wastes.

Fill-in Key:

1. Superfund
2. EPA
3. hazardous
4. incineration
5. decomposition
6. permanent
7. phytoremediation
8. 60
9. sanitary
10. methane
11. waste-to-energy
12. reusing
13. recycling
14. recycling
15. composting
16. compost
17. co-composting
18. externality
19. sewage
20. radioactive

Multiple Choice Questions

1. Love Canal was a hazardous-waste disposal site in _____.
   a. Brazil
   b. Woburn, Massachusetts
   c. Denver, Colorado
   * d. Niagara Falls, New York
   e. Canada

2. Improper hazardous waste disposal has which of the following costly effects?
   a. groundwater contamination
   b. habitat destruction
   c. fish kills
   d. soil contamination
   * e. all of the above

3. The U.S. Office of Technology Assessment estimate that it will cost _____ billion to clean up the 10,000 sites in the U.S. that pose a risk to health.
   a. $20
   b. $40
   c. $60
   d. $80
   * e. $100

4. This law created the *Superfund*, a fund financed by state and federal governments and by taxes on chemical and oil companies:
   a. Resource Conservation and Recovery Act
   * b. Comprehensive Environmental Response, Compensation and Liability Act
   c. Occupational Safety and Health Act
   d. Toxic Substances Control Act
   e. Oil Pollution Act

5. In 1995, the EPA listed _____ sites on the National Priority List because of their potential health threat.
   a. 225
   b. 425
   c. 625
   d. 825
   * e. 1225

6. Leaking underground storage tanks are a major environmental problem since leaks can:
   a. reduce the profits for the company owning the tanks.
   * b. contaminate groundwater.
   c. cause an unsightly mess.
   d. go undetected for some time.
   e. be cause a nuisance because of the odor problem.

7. About ___% of the cleanup and replacement of leaking underground storage tanks is being financed and performed by private industry.
   a. 20
   b. 40
   c. 60
   d. 80
   * e. 90

8. Which of the following responses is a weakness in the current Resource Conservation and Recovery Act?
   a. Municipal waste contain toxic chemicals and is not classified as a hazardous waste.
   b. Sewage and untreated wastewater by can contain toxic chemicals and is not considered to be hazardous waste.
   c. Agricultural wastes are not regulated.
   d. Mill and mind tailing are excluded from the law.
   * e. all of the above

9. 13 European nations and 113 non-European nations have signed an agreement that bans the transfer of hazardous wastes to developing nations; this agreement is the:
   a. Montreal Protocol
   b. Geneva Accord
   * c. Basel Convention
   d. Helsinki Convention
   e. Earth Summit

10. Which of the following responses is a problem with the current operation of the *Superfund*?
    a. The cleanup has proven to be very costly.
    b. The cleanup is slow.
    c. Most of the money is going to legal fees.
    d. It has created a legal nightmare.
*   e. all of the above

11. This law is designed to eliminate illegal and improper waste disposal:
    a. Toxic Substances Control Act
    b. Comprehensive Environmental Response, Compensation and Liability Act
*   c. Resource Conservation and Recovery Act
    d. Oil Pollution Act
    e. Occupational Safety and Health Act

12. Leaking underground storage tanks are covered under the 1984 amendments of this law:
    a. Toxic Substances Control Act
    b. Oil Pollution Act
*   c. Resource Conservation and Recovery Act
    d. Comprehensive Environmental Response, Compensation and Liability Act
    e. Occupational Safety and Health Act

13. Burning hazardous wastes at high temperature is called:
*   a. incineration
    b. substitution
    c. detoxification
    d. decomposition
    e. land disposal

14. This technique of hazardous waste treatment uses high pressure and temperatures of 840-1100°F to break down compounds into smaller, biodegradable molecules:
    a. incineration
    b. substitution
    c. biofiltration
*   d. low-temperature decomposition
    e. detoxification

15. Process manipulation, substitution, and reuse are _____ that reduce or eliminate hazardous waste production in industry.
    a. perpetual storage options
*   b. in-plant options
    c. incineration options
    d. conversion options
    e. landfill options

16. Mixing hazardous wastes with soil to encourage the degradation of toxic materials is called:
    a. recycling
*   b. land disposal
    c. substitution
    d. incineration
    e. thermal treatment

17. Cleaning up contaminated waste sites using plants is referred to as:
    a. composting
    b. substitution
*   c. phytoremediation
    d. restoration
    e. biofiltration

18. Americans produce approximately _____ pounds of solid waste per year.
    a. 50
    b. 500
    c. 1,000
    d. 1,500
*   e. 2,500

19. The largest fraction of U.S. municipal solid waste by weight is:
    a. plastics
    b. wood
    c. yard wastes
*   d. paper and paperboard
    e. food wastes

20. Excavated pits lined by impermeable synthetic liners and thick impermeable layers of clay are:
* a. secured landfills
  b. dumps
  c. surface impoundment
  d. warehouses
  e. open pits

21. This law established a timetable for the Department of Energy to select a deep underground disposal site for high-level radioactive wastes:
  a. Resource Conservation and Recovery Act
  b. Toxic Substance Control Act
* c. Nuclear Waste Policy Act
  d. Occupational Safety and Health Act
  e. Oil Pollution Act

22. Which of the following responses are personal actions which can reduce hazardous waste?
  a. Recycle oil from your automobile.
  b. Use nontoxic, water-soluble finishes.
  c. Recycle and compost all the waste you can.
  d. Buy recycled goods.
* e. all of the above

23. Reducing the amount of solid wastes by recycling and reuse are called the:
  a. output approach
  b. input approach
* c. throughput approach
  d. end-of -pipe
  e. green revolution

24. A natural or manmade depression into which solid wastes are dumped, compressed, and covered daily with a layer of dirt is a/an:
  a. garbage dump
  b. open pit
* c. sanitary landfill
  d. recycling center
  e. hazardous waste site

25. Which of the following responses are factors that influence the volume of solid waste in the U.S.?
  a. population size
  b. per capita consumption
  c. product durability
  d. reuse and recycling rates
* e. all of the above

26. The traditional approach to solid waste management consists of ways to deal with the trash produced, such as landfilling or incineration; this strategy is called:
* a. output approach
  b. input approach
  c. throughput approach
  d. reuse
  e. recycling

27. Activities that reduce the amount of waste by lowering consumption and increasing product durability are called the:
  a. output approach
* b. input approach
  c. throughput approach
  d. end-of-pipe
  e. recycling

28. Incineration accounts for about ___% of solid waste disposal in the United States.
  a. 5
  b. 10
* c. 15
  d. 20
  e. 25

29. Incineration of garbage can be used to produce heat and electricity and incinerators are sometimes called:
  a. heat plants
  b. energy plants
* c. waste to energy plants
  d. power plants
  e. cogeneration plants

30. Which of the following responses is a problem associated with landfills?
    a. Usable land is in short supply or is expensive for landfills.
    b. Landfills are energy intensive.
    c. Toxic substances disposed in landfills can pollute groundwater.
    d. Landfills produce methane gas which is explosive.
*   e. all of the above

31. The long term goal of this law, passed by Congress in 1972, was to phase out all ocean dumping:
    a. Resource Conservation and Recovery Act
*   b. Marine Protection, Research and Sanctuaries Act
    c. Ocean Dumping Ban Act
    d. Oil Pollution Act
    e. Toxic Substance Control Act

32. This law, passed by Congress in 1988, called for an end to sewage sludge disposal at sea by 1991:
    a. Marine Protection, Research and Sanctuaries Act
*   b. Ocean Dumping Ban Act
    c. Toxic Substances Control Act
    d. Oil Pollution Act
    e. Resource Conservation and Recovery Act

33. A cost that is passed on to the public from pollution and other harmful effects of these activities is called a/an:
    a. tariff
*   b. economic externality
    c. tax
    d. levy
    e. economic incentive

34. Local, state and federal government spending amount to ___% of the total Gross National Product.
    a. 10
*   b. 20
    c. 30
    d. 40
    e. 50

35. Techniques that reduce the amount of waste entering the waste stream are called:
    a. waste management
*   b. source reduction
    c. end-of-pipe
    d. recycling
    e. reuse

36. The amount of waste entering the waste stream can be reduced by:
    a. Buying fewer goods.
    b. Buy the most durable product available.
    c. Make products smaller and reduce packaging.
    d. a and b only
*   e. a, b, and c

37. The return of materials to manufacturers where they can be melted down or refashioned into new products is called;
    a. restoration   d. reuse
*   b. recycling   e. renewal
    c. composting

38. Returning the nutrients of organic matter such as yard waste and food scraps to the soil is called:
    a. reuse
    b. recycling
*   c. composting
    d. restoration
    e. waste removal

39. Nationwide, organic wastes such as leaves, grass clippings, and kitchen scraps constitute about ___% of the garbage dumped into landfills or incinerated.
    a. 10   d. 40
*   b. 20   e. 50
    c. 30

40. Combining sewer sludge with municipal solid waste materials for composting is called:
*   a. co-composting
    b. cogeneration
    c. recycling
    d. biofiltration
    e. detoxification

25

Sustainable Ethics: The Foundation of a Sustainable Society

Chapter Outline

The Frontier Mentality Revisited

Sustainable Ethics: Making the Transition
   Leopold's Land Ethic: Planting the Seed

A New View to Meet Today's Challenges: Sustainable Ethics
   Toward a Humane, Sustainable Future

Developing and Implementing Sustainable Ethics
   Promoting Models of Sustainability
   Education
   Churches
   Declarations of Sustainable Ethics and Policy
   A World Organization Dedicated to Sustainable Development
   A Role for Everyone

Overcoming Obstacles to Sustainability
   Faith in Technological Fixes
   Apathy, Powerlessness, and Despair
   The Self-Centered View
   Ego Gratification
   Economic Self-Interest and Outmoded Governmental Policies

Sustainable Ethics: How Useful Are They?

Key Terms

ethics	frontier ethic	sustainable ethics
land ethic	technological optimism	utilitarianism
natural rights		

Objectives

1. Discuss the "frontier ethic", trace its development and summarize its environmental consequences.
2. Compare the "frontier ethics" to "sustainable ethics" and discuss the how "sustainable ethics" can be promoted.
3. List some obstacles to sustainability and how these may be overcome.
4. Compare and contrast utilitarianism, divine law and natural rights theories as foundations for establishing environmental ethics.
5. Discuss the importance of Aldo Leopold's "land ethic" in the development of environmental ethics.

Lecture Outline

I. The Frontier Mentality Revisited
    A. This is the dominant paradigm of modern society.
    B. The frontier paradigm has three precepts: the world's resource base and resiliency are limitless, humans are apart from nature, and nature is a rival to be conquered.
    C. This mentality, socialized and politicized serves to justify the individual's pursuit of personal prosperity at the expense of ecosystem stability and diversity.

II. Sustainable Ethics: Making the Transition
    A. Leopold's Land Ethic: Planting the Seed
       1. Aldo Leopold's *A Sand County Almanac* first outlined and made a plea for a land ethic. Though valuable as an ideal goal, Leopold's land ethic offered little guidance for making the switch to a sustainable society.

III. A New View to Meet Today's Challenges: Sustainable Ethics
    A. Sustainable ethics embrace the reverse of the tenets of the frontier society, it is based on conservation reuse and cycling, renewable resources, and population control, and replaces human arrogance toward nature with respect and restraint.
    B. The sustainable ethics go beyond the land ethic by giving shape to Leopold's ideals.
    C. The emphasis of sustainable ethics is on restraint, cooperation, and obligation was reflected in the 1992 Rio conference, an indication that these ideas may be beginning to take hold.

IV. Developing and Implementing Sustainable Ethics
    A. Promoting Models of Sustainability
       1. By seeking to spread awareness of working models of sustainability, we may both inspire and guide others on the path to sustainability.
    B. Education
       1. An interdisciplinary, holistic approach to education at all levels will facilitate the transition to a sustainable ethic-based society.
    C. Religious Organizations
       1. Increased efforts on the part of religious leaders may promote environmental awareness among parishioners.

D. Declarations of Sustainable Ethics and Policy
   1. The ratification of constitutional amendments can help nations send a strong message of environmental commitment to the world.
E. A World Organization Dedicated to Sustainable Development
   1. Through support of an international organization functioning as an environmental watchdog, nations may successfully unite in efforts to address global environmental issues.
F. A Role for Everyone
   1. Each one of us can play a significant role in bringing about the needed changes in collective mindset. While no one of us can change the world, we can each make changes which minimize our and our families' environmental impacts, and which serve as a model for others to follow.

V. Overcoming Obstacles to Sustainability
   A. Faith in Technological Fixes
      1. Our society has a nearly unshakable faith that technology can fix whatever we damage; such optimism is usually at odds with the facts and with critical thinking.
   B. Apathy, Powerlessness, and Despair
      1. Fostered by conditioning, a sense of powerlessness, and feelings of insignificance, apathy causes us to shirk individual responsibility for environmental protection; this leads to the paradox of inconsequence.
   C. The Self-Centered View
      1. An excessively self-centered approach to life encourages consumerism and pursuit of personal material well-being at considerable environmental expense; it discourages personal involvement and sacrifice for environmental quality.
   D. Ego Gratification
      1. Materialism and overconsumption may have deep psychological roots in the development of personal and family relationships.
   E. Economic Self-Interest and Outmoded Governmental Policies
      1. Revision of most countries' socioeconomic systems in keeping with the precepts of sustainability will be necessary if root causes of our environmental crisis are to be addressed.

VI. Sustainable Ethics: How Useful Are They?
   A. Value Judgments and Decision Making
      1. Our values strongly influence our decision-making, and the decisions we make profoundly affect the environment.
      2. Examples of value systems within which decisions are made include utilitarianism, divine law, and natural rights.

Suggestions for Presenting the Chapter

- Instructors should challenge the students to think about their values in reference to the "frontier ethic". The process of clarifying values should be encouraged throughout this course. Discussion groups are good tools to help students become aware of their values and think about change.

- The topic of cost and external costs should be stressed. Students should understand how to be more sustainable in their consumer activities.
- Assign some readings from texts on environmental ethics or other relevant works.
- Recommended web sites:

United Nations Sustainable Development/Agenda 21:
> http://www.un.org/esa/sustdev/agenda21.htm

United Nations Division for Sustainable Development: http://www.un.org/esa/sustdev/

John Muir Exhibit/Sierra Club: http://www.sierraclub.org/john_muir_exhibit/

Excerpts from the Works of Aldo Leopold:
> http://gargravarr.cc.utexas.edu/chrisj/leopold-quotes.html

Deep Ecology/Arne Naess:
> http://home.clara.net/clara.net/h/e/u/heureka/webspace/gaia/deep-eco.htm

True/False Questions

1. ___ The Earth is an unlimited supply of resources for exclusive human use.
2. ___ Humans are not subject to the laws of nature.
3. ___ Human success is best achieved by controlling nature.
4. ___ The frontier ethic is unique to the Western World.
5. ___ Aldo Leopold described a land ethic in his book, *A Sand County Almanac.*
6. ___ Participation by all sectors of society is essential to building a sustainable society.
7. ___ Technology can solve all the world's problems, including any environmental damage we do on the Earth.
8. ___ Unsustainable economic systems and governments that support them are major barriers to sustainable development.
9. ___ Utilitarianism is the belief that the worth of anything is determined by its usefulness.
10. ___ The idea that all living things have rights, irrespective of their value to human society is known as natural rights.
11. ___ Values affect the way we act.
12. ___ Humans depend on natural systems for their survival.
13. ___ Apathy and feelings of powerlessness and despair hinder progress toward sustainability.
14. ___ Concern for the self and blame shifting hinder efforts to foster individual action and responsibility.
15. ___ The principle of fairness to future generations is known as intragenerational equity.
16. ___ Churches can become an important force in creating global change needed to foster a sustainable future.
17. ___ Publicizing models of sustainable action offers inspiration and practical examples of what individuals and groups can do to build a sustainable future.
18. ___ Participation by all sectors of society is essential to building a sustainable society.
19. ___ Feelings of inadequacy are often offset by the accumulation of material possessions.

20.___ A dollar invested in energy conservation reduces carbon dioxide emissions seven times more than a dollar invested in nuclear power.

True/False Key:

1. F 2. F 3. F 4. F 5. T 6. T 7. F 8. T 9. T 10. T 11. T 12. T 13. T 14. T 15. F 16. T 17. T 18. T 19. T 20. T

Fill-in-the-Blank Questions

1. _____ are a part of nature.
2. Humans depend on _____ systems for their survival.
3. _____ _____ wrote *A Sand County Almanac*.
4. The _____ ethic is the idea that humans are a part of a larger community that includes soils, water, plants and animals.
5. The found of the Sierra Club was _____ _____.
6. _____ ethics suggests that humans are a part of nature.
7. The United Nations Commission on Sustainable Development is an _____ organization dedicated to promoting sustainable development.
8. The belief that _____ can solve all the world's problems hinders the development of a sustainable ethic.
9. The development of our own sense of the self, our own separate identity, is known as the _____ self.
10. Unsustainable _____ systems and governments that support them are barriers to sustainable development.
11. Feelings of inadequacy are often offset by the accumulation of _____ possessions.
12. Placing the responsibility for environmental damage on other people is called _____ shifting by the author of your textbook.
13. The idea that the Earth is an unlimited supply of resources for exclusive human use is part of the _____ ethic.
14. The main tenet of sustainable ethics is "there is not always ____".
15. The ever-increasing production and consumption in a world of limits could destroy the ____-_____ systems of the planet.
16. The core value of sustainable ethics is respect and care for the _____ of life.
17. _____ equity refers to the responsibility we have toward future generations.
18. Fairness to other species is known as _____ justice.
19. Fairness to others who are alive today is known as intragenerational _____.
20. The giving of money in lieu of action, thought or care has been called "cash _____" by the author of your textbook.

Fill-in Key:

1. humans
2. natural
3. Ado Leopard
4. land
5. John Moiré
6. sustainable
7. international
8. technology
9. derived
10. economic
11. material
12. blame
13. frontier
14. more
15. life-support
16. community
17. intergenerational
18. ecological
19. equity
20. conscience

Multiple Choice Questions

1. Which of the following responses is **not** part of the frontier ethic?
   a. Economics is all that matters.
   b. New laws and technologies will solve our problems.
   * c. We must understand and cooperate with nature.
   d. When the supply runs out, move elsewhere.
   e. The earth is an unlimited bank of resources for exclusive human use.

2. *A Sand County Almanac* was written by:
   a. John Muir
   * b. Aldo Leopold
   c. Ralph Waldo Emerson
   d. Henry Thoreau
   e. Bob Marshall

3. The idea that humans are part of a larger community that included the land is called:
   * a. the land ethic.
   b. the frontier ethic.
   c. the green ethic.
   d. manifest destiny.
   e. maximum sustainable yield.

4. Which of the following responses is a way to encourage the development of sustainable ethics?
   a. Publicizing models of sustainable action.
   b. All levels of education can foster transmission of sustainable thinking.
   c. Churches can become an important force in promoting a new Earth ethic.
   d. States, nations and groups of nations can promote adoption of sustainable ethics.
   * e. all of the above

5. In 1992, over 170 nations adopted the
   _____, a kind of international declaration
   for sustainable development.
   a. Helsinki Accord
   b. Geneva Convention
   * c. Rio Declaration
   d. Montreal Protocol
   e. Earth Summit

6. The UN Commission on Sustainable
   Development (UNCSD) was created at the:
   a. 1997 Rio Conference
   * b. 1992 Earth Summit
   c. Helsinki Commission
   d. Montreal Accord
   e. London Agreement

7. Which of the following responses is an
   obstacle to sustainability?
   a. Faith in technological fixes.
   b. Apathy and feelings of powerlessness.
   c. Concern for the self.
   d. Ego gratification.
   * e. all of the above.

8. The doctrine by which the worth of anything
   is determined by its usefulness is called:
   a. natural rights
   * b. utilitarianism
   c. nonanthropocentric ethics
   d. ecocentrism
   e. misanthropic

9. The doctrine which suggests that all living
   things have rights irrespective of their value
   to human society is called:
   * a. natural rights
   b. utilitarianism
   c. anthropocentric ethics
   d. humanism
   e. ecohumanism

10. The founder of the Sierra Club was:
    * a. John Muir
    b. Aldo Leopold
    c. Henry Thoreau
    d. Theodore Roosevelt
    e. Bob Marshall

11. The idea that we are subject to the laws that
    govern all life on Earth, not immune to them
    is a principle is called:
    a. freedom
    b. natural rights
    * c. dependence
    d. independence
    e. utilitarianism

12. Which of the following responses is a
    principle of sustainability?
    a. recycling
    b. restoration
    c. conservation
    d. renewable resources
    * e. all of the above

13. Promoting models of sustainability can be
    done by:
    a. books
    b. news reports
    c. newspaper articles
    d. government reports
    * e. all of the above

14. Which of the following responses is a reason
    why we should feel responsible for future
    generations?
    a. Future generations will be the same as we
       are.
    b. We have no more rights over the world
       and its resources than anyone else.
    c. Even after we die, the effects of our life
       continue.
    d. Our survival as a species is more
       important than our individual survival.
    * e. all of the above

15. The Rio Declaration is a statement of 27
    principles regarding:
    a. oil pollution at sea.
    * b. environment and development.
    c. the Endangered Species Act.
    d. international pesticide pollution.
    e. the Montreal Protocol

16. Many people express extreme optimism in the power of technology to solve our problems. This belief is sometimes called:
    a. humanism
    b. socialism
  * c. technological optimism
    d. social Darwinism
    e. capitalism

17. Author and social critic, _____, coined the phrase, "the me-generation" to describe the Americans of the 1980s who seemed intent on self indulgence.
  * a. Tom Wolfe
    b. Paul Ehrlich
    c. Edward Abbey
    d. Dave Foreman
    e. Carl Sagan

18. As we pass from infancy to adulthood we develop a sense of the self, of our own separate identity, this identity is called the:
  * a. derived self
    b. little child
    c. material self
    d. transcendent self
    e. natural self

19. _____ may be one of the most useful sets of values because it serves people, the environment, and the economy.
    a. Frontier ethics
    b. Humanism
  * c. Sustainable ethics
    d. Utilitarianism
    e. Socialism

20. _____ is the yardstick of many utilitarian decisions.
    a. Environmental damage
  * b. Economics
    c. Individual welfare
    d. Animal rights
    e. Animal welfare

26

Sustainable Economics: Understanding the Economy and Challenges Facing the Industrial Nations

Chapter Outline

Economics, Environment, and Sustainability
Economic Systems
The Law of Supply an Demand
Environmental Implications of Supply and Demand
Measuring Economic Success: The GNP

The Economics of Pollution Control
Cost-Benefit Analysis and Pollution Control
Who Should Pay for Pollution Control
Does Pollution Control Always Cost Money?

The Economics of Resource Management
Time Preference
Opportunity Cost
Discount Rates
Ethics

What's Wrong with Economics: An Ecological Perspective
Economic Shortsightedness
Obsession with Growth
Exploitation of People and Nature

Creating a Sustainable Economic System: Challenges in the Industrial World
Harnessing Market Forces to Protect the Environment
Corporate Reform: Greening the Corporation
Green Products and Green Seals of Approval
Appropriate Technology and Sustainable Economic Development
A Hopeful Future

Environmental Protection vs. Jobs: Problem or Opportunity?

Key Terms

economics
descriptive economics
sustainable economics
supply
free market
cost-benefit analysis
taxpayer-pays option
repercussion costs
discount rates
full-cost pricing
marketable permit
appropriate technology

inputs
normative economics
command economies
demand
gross national product
law of diminishing returns
direct costs
time preference
ethics
net economic welfare
green products

outputs
ecological economics
market economies
market price equilibrium
economic externality
consumer-pays option
indirect costs
opportunity cost
green taxes
sustainable economic welfare index
environmental auditing

Objectives

1. Define the following terms: "economics", "descriptive economics", "normative economics", "ecological economics", and "sustainable economics".
2. List the two types of economic systems and their characteristics.
3. Discuss the law of supply and demand and the environmental implications of this law.
4. Discuss how the following concepts fit into understanding the economics of pollution control: "economic externality", "cost-benefit analysis", "law of diminishing returns", "direct costs", "indirect costs" and "repercussion costs".
5. Define the following terms and explain how each term relates resource management: "time preference", "opportunity cost", "discount rates", and "ethics".
6. Discuss some reasons current market economics lead to unsustainable activities.
7. Discuss some alternative measure of economic progress other than the gross national product (GNP).
8. Summarize the changes in our economic system that would foster environmental sustainability.
9. List some characteristics of a sustainable economy.
10. Define the term "appropriate technologies" and list the characteristics of an appropriate technology.
11. Discuss how environmental protection could be a stimulant to economic progress.

Lecture Outline

I. Economics, Environment, and Sustainability - Economics is the study of the production, distribution, and consumption of goods and services. Both descriptive and normative economics must be employed in any study of the relationship between economics and the environment. Sustainable economics is that branch of economics which integrates normative and descriptive approaches in the service of both people and the environment.
    A. Economic Systems
       1. Economics seeks to answer three questions: what, how, and for whom commodities should be produced.

    2.   These questions are answered differently in command and market economies; most nations' economies are mixed, with elements of both types integrated in them.

B.   The Law of Supply and Demand

    1.   Price is the main governor of behavior in a market economy; it is largely determined by supply and demand.

    2.   The market price equilibrium is the point at which supply and demand curves graphically intersect, and represents a price compromise between producers and consumers.

    3.   The law of supply and demand can have serious implications for the environment and may work against conservation efforts, as it fails to take into account the finite nature of many resources.

C.   Environmental Implications of Supply and Demand

    1.   Supply and demand can spawn a variety of wasteful and unsustainable practices.

D.   Measuring Economic Success: The GNP

    1.   Gross national product (GNP) is the most widely used measure of a nation's economic activity.

II. The Economics of Pollution Control - Traditionally, most economies have regarded pollution as an economic externality. In response to citizen complaint, some governments established pollution control standards which internalized the costs of industrial pollution.

A.   Cost-Benefit Analysis and Pollution Control

    1.   The goal of cost-benefit analysis is to maximize pollution control at minimum cost.

    2.   Problems with cost-benefit analysis include the impossibility of quantifying certain costs and benefits, such as human life, pain and suffering, and aesthetic/intrinsic values; if not quantified, they cannot easily enter into the analysis.

B.   Who Should Pay for Pollution Control?

    1.   Either the consumer, the taxpayer, or both must bear the cost of pollution abatement programs; determining who should pay in a given situation is difficult and often controversial.

    2.   Note that pollution control does not always cost but often pays the polluter, when all direct, indirect, and repercussive costs of pollution are figured in.

III. The Economics of Resource Management - Economic considerations often influence our behavior and decisions regarding natural resources and pollution.

A.   Time Preference

    1.   Time preference refers to one's willingness to postpone a certain reward today for a greater reward in the future; it is influenced by current needs, uncertainty outcome, inflation rates, and the rate of return on the postponed reward.

    2.   With regard to natural resources, differing time preferences may lead to either depletion or conservation strategies.

B.   Opportunity Cost

    1.   This is the cost of lost opportunities; when high, it may discourage the adoption of conservation strategies, though the reverse is increasingly the case.

C.   Discount Rates

       1.   The discount rate is a decision-making tool reflecting time preference and opportunity costs; its application is often used to justify the rapid depletion of natural resources.
  D. Ethics
       1.   Ethical considerations are noneconomic factors which often affect our economic decisions.

IV. What's Wrong with Economics: An Ecological Perspective
  A. Economic Shortsightedness
       1.   Vision is needed to adjust supply and demand economics to reflect ecological realities.
  B. Obsession with Growth
       1.   The major bias of descriptive and normative economics is the tenet that economic growth is always desirable; this tenet rests upon the frontier belief that there is always more of everything needed to fuel such growth.
       2.   The doctrine of growth requires an ever-expanding population and ever-increasing per capita consumption, both environmentally disastrous.
  C.   Growth and the GNP: Some Fundamental Flaws
       1.   The GNP is value-neutral and counts wasteful or remedial expenditures as well as those genuinely contributory to standard of living improvement.
       2.   Subtracting these disamenities from the GNP yields net economic welfare (NEW), a more accurate measure of an economy's service to society. As pollution and congestion increase, the disparity between GNP and NEW increase.
       3.   Since GNP inherently favors growth and ignores accumulated wealth, it cannot guide or reflect the transition to a sustainable economy; similarly, by quantifying health statistics but ignoring total well-being, we fail to accurately measure a country's overall condition.
  D. Ending Our Obsession with Growth
       1.   New, more holistic and realistic measures of economic success are needed to supercede our current obsession with growth as the measure of prosperity.
  E. Making the Economic System Work for Us
       1.   Rather than fatalistically accepting the divergence between GNP and NEW, a sustainable society must have as its goal the elimination of this discrepancy.
  F. Rethinking Growth: Focusing on Development
       1.   Limited economic growth which focuses on sustainable forms of development is an environmentally preferable alternative to the current pattern of unsustainable growth.
  G. The Economic System and Dependency
       1.   Economic interdependency has fostered unsustainability in many regions of the world. Regionalism and localism are more inherently stable and ecologically appropriate.
  H. Exploitation of People and Nature
       1.   The traditional economy inherently exploits both people and nature

V. Creating a Sustainable Economic System: Challenges in the Industrial World
  A. Harnessing Market Forces to Protect the Environment - Some have proposed that the government harness existing marketplace forces in service of environmental protection.
       1.   Economic Disincentives

      a.   Through user fees and pollution taxes, economic disincentives can make environmental protection more attractive to individuals and companies than continued environmental abuse.

   2. Economic Incentives

      a.   Incentives such as tax credits and economic grants can insure compliance with legislation aimed at protecting the environment.

   3. The Permit System

      a.   While not without critics, due to its ethical implications, marketable permits may be used to regulate total pollution emissions in a given region while minimizing costs of compliance.

   4. Removing Market Barriers

      a.   Legislative and regulatory reform are needed to make environmentally-friendly practices more feasible.

B. Corporate Reform: Greening the Corporation

   1. Many corporations are beginning to reform their operating principles to incorporate guidelines for responsibility such as the Coalition for Environmentally Responsible Economies Principles (CERES)

C. Green Products and Green Seals of Approval

   1. Corporate environmentalism is increasingly reaching the consumer in the form of environmentally friendly products; these can often be identified by the Green Seal or Green Cross mark of approval.

D. Appropriate Technology and Sustainable Economic Development

   1. An integral element of sustainability is the development and use of appropriate technology, that is technology which is efficient and proportionately scaled to the task at hand.

E. A Hopeful Future

   1. While progress towards sustainability is unsatisfactorily slow, we should remain optimistic that the trends of unsuitability will be changed as awareness and concern increase worldwide.

VI. Environmental Protection vs. Jobs: Problem or Opportunity?

A. Nearly all objective studies and analyses show that this is a false dichotomy; environmental protection, in the long run, is always more compatible with a healthy economy and full employment than is business as usual.

Suggestions for Presenting the Chapter

- Instructors should stress the problems with current economic theory and how they can be improved.
- Economic activity and environmental health are not mutually exclusive. Instructors should explain how local and regional economies can become more sustainable.
- Students should be encouraged to identify sustainable economic activity in your area/region. Discussion of local economies is a good way to illustrate the important points of the chapter.
- A trip to a sustainable business is a good way to promote sustainable activity in the community.
- Recommended web sites:

The World Bank: http://www.worldbank.org/
The Development Center for Appropriate Technology: http://www.azstarnet.com/~dcat/
World Trade Organization: http://www.wto.org/wto/
Worldwatch Institute: http://www.worldwatch.org/index.html
President's Council on Sustainable Development/Sustainable Communities Task Force:
    http://www.whitehouse.gov/PCSD/Publications/suscomm/ind_suscom.html

True/False Questions

1. ___ Economics is the study of the production, distribution and consumption of goods and services.
2. ___ Normative economics concerns itself with supplying the needs of people while protecting the environment.
3. ___ Two types of economic systems exist: command and market economies.
4. ___ Governments dictate production and distribution goals in command economies.
5. ___ The principle that most things that people want are limited is the law of scarcity.
6. ___ The predictable interplay of price, supply and demand constitutes the law of supply and demand.
7. ___ Supply refers to the amount of goods people want.
8. ___ Economies where supply and demand is totally controlled by market forces is a free market system.
9. ___ The gross national product is the market value of the nation's output.
10. ___ A cost to society and the environment not paid directly are called direct costs.
11. ___ Repercussion costs are incurred by the industry due to the image problems arising from an oil spill or other environmental damage.
12. ___ The measure of one's willingness to postpone some current income for greater returns in the future is called opportunity cost.
13. ___ Opportunity cost is the cost of lost opportunities.
14. ___ Levies on environmentally undesirable products or activities are called green taxes.
15. ___ An economic system that seeks only to meet human needs and ignores environmental concerns is called a sustainable economy.
16. ___ Appropriate technologies are inefficient, rely on foreign resources and produce considerable amounts of pollution.
17. ___ The gas-guzzler tax is a tax on cars that get poor gas mileage.
18. ___ Grants and tax incentives can be used by governments to encourage sustainable business practices and products.
19. ___ Marketable permits for emissions provide an economic incentive for companies to reduce pollution.
20. ___ Greater local and regional self-reliance may be essential to achieving a sustainable future.

Fill-in-the-Blank Questions

1. Economic systems tend to _____ people and the environment.
2. One characteristic of a sustainable economy is that it uses all _____ efficiently.
3. Sustainable economies promote _____ self-reliance
4. _____ technologies rely on local resources, are efficient and produce little if any pollution.
5. _____ products are environmentally friendly goods and services.
6. _____ is the dollar value of goods per hour of paid employment.
7. A problem with the GNP is that it fails to account for the destruction of _____ assets.
8. Economic growth is fueled by _____ growth and ever increasing per capita consumption.
9. Soil, the _____ layer, and current climate are vital Earth assets.
10. Economic systems often fail to take into account long-term supplies, a trend that results in an _____ of many natural resources.
11. The _____ rate allows investors and economists to determine the present value of different profit-making options.
12. _____ cost is the cost of lost opportunities.
13. A measure of one's willingness to postpone some current income for greater returns in the future is known as _____ preference.
14. The Exxon Valdez oil spill in Alaska happened in _____.
15. Indirect costs due to the Exxon Valdez oil spill were paid by the state and _____ agencies.
16. The law of _____ returns predicts that for each additional dollar invested there will be a smaller return.
17. A cost to society and the environment that is hidden and not paid directly is known as an _____ externality.
18. Economics is the study of _____, distribution and consumption of goods and services.
19. The intersection of the supply and demand curves is known as the market price _____.
20. _____ refers to the amount of a resource, product or service that is available.

3. regional
4. appropriate
5. Green
6. productivity
7. natural
8. population
9. ozone
10. underpricing
11. discount
12. opportunity
13. time
14. 1989
15. federal
16. diminishing
17. economic
18. production
19. equilibrium
20. supply

Multiple Choice Questions

1. The study of the production, distribution and consumption of goods and services is called:
   a. sociology
   b. anthropology
   * c. economics
   d. ecology
   e. psychology

2. Command economies are largely run by:
   a. individuals
   b. small companies
   c. large corporations
   d. trade organizations
   * e. governments

3. A country with a command economy would be:
   a. United States
   b. Great Britain
   c. Germany
   * d. Cuba
   e. Canada

4. Economies which produce goods and services for which there is the highest demand are called:
   a. demand economies
   b. command economies
   * c. market economies
   d. inflated
   e. saturated

5. The law of scarcity states that:
   * a. most things people want are limited.
   b. price is not a rationing mechanism.
   c. you cannot have what you want.
   d. scarce items will get scarcer.
   e. the harder to find something is the lower the price.

6. Prices in a market economy are largely determined by the interaction of supply and

   _____.
   a. cost of production
   b. surpluses
   c. time of purchase
   * d. demand
   e. inventory

7. The price at which consumers are willing to buy a product and at which producers can afford to produce it is the:
   a. point of no return
   b. maximum sustainable yield
   * c. market price equilibrium
   d. markup
   e. discount rate

8. The Timber, Fish and Wildlife agreement established a new type of management process called adaptive management. This TFW agreement occurred in which of the following states?
   a . Idaho
   b. Montana
   * c. Washington
   d. Oregon
   e. California

9. A measure of the economic output of the nation, including all goods and services is the:
   * a. gross national product
   b. economic product
   c. gross income product
   d. cost benefit analysis
   e. cost of goods and services

10. The costs of pollution control that companies incur are _____ and passed on to the consumer in the form of a price increase.
   a. externalized
   * b. internalized
   c. deducted
   d. avoided
   e. standardized

11. The advantage of sustainable approaches to reducing pollution over end-of-pipe controls is that they can:
   * a. reduce emissions at a far lower cost.
   b. take a longer time.
   c. are about equally effective as conventional methods.
   d. require large amounts of new technology
   e. will require complete retooling of industry.

12. If industry passes the cost of pollution control on to the consumer, this strategy is known as the:
   * a. consumer-pays option
   b. taxpayer-pays option
   c. government-pays option
   d. market forces option
   e. free market option

13. Government payments to coal miners who suffer from black lung disease as a result of exposure to coal dust, is called the:
   a. consumer-pays option
   * b. taxpayer-pays option
   c. insurance option
   d. no-fault option
   e. socialism option

14. Exxon paid expenses to clean up the oil spill in the weeks and months after the 1989 Exxon Valdez spill. These costs are called:
   * a. direct costs
   b. indirect costs
   c. repercussion costs
   d. overhead
   e. operating expenses

15. A cost to society and the environment not paid directly by manufacturer or consumer is a/an:
   a. tax
   * b. externality
   c. rebate
   d. penalty
   e. surcharge

16. Costs incurred by a state or federal agency as the result of an oil spill are called:
   a. direct costs
   * b. indirect costs
   c. repercussion costs
   d. overhead
   e. operating costs

17. Boycott of the Exxon company after the oil spill cost the company revenue. This cost is called the:
    a. direct costs
    b. indirect costs
*   c. repercussion costs
    d. overhead
    e. operating costs

18. A measure of willingness to postpone some current income for greater returns in the future is called the:
    a. time delay
*   b. time preference
    c. repercussion cost
    d. cost of waiting
    e. future cost

19. An economic index that reflects one's time preference, opportunity costs, and allows investors to determine the economic value of different profit making options is the:
*   a. discount rate
    b. down payment
    c. gross income
    d. interest rate
    e. instantaneous growth rate

20. If a company incorporates all the costs of producing a product or service , including environmental externalities, this is called:
    a. user fees
*   b. full-cost pricing
    c. discount pricing
    d. retailing
    e. price gouging

21. Economic growth is fueled by:
    a. population growth.
    b. increasing per capita consumption.
    c. a notion of endless resources.
    d. the idea that growth is progress.
*   e. all of the above

22. The Gross National Product fails to take in consideration:
    a. resource depletion
    b. accumulated wealth
    c. the distribution of wealth
    d. economic activity that reduces the quality of life
*   e. all of the above

23. _____ of American children live in poverty.
    a. one-eighth
*   b. one-fifth
    c. one-fourth
    d. one-third
    e. one-half

24. Nearly ___ million Americans have no health insurance.
    a. 10
    b. 20
    c. 30
*   d. 40
    e. 50

25. Kenneth Boulding coined the term *cowboy economy* and suggested that it needs to be replaced by a _____ economy.
    a. mother earth
    b. frontier
*   c. spaceship
    d. green
    e. natural

26. Taxes on raw materials, paid by producers and ultimately passed on to consumers, such as the severance tax paid by coal companies, is called a:
*   a. green tax
    b. surcharge
    c. rebate
    d. opportunity cost
    e. externality

27. Yale economists, Nordhaus and Tobin, have created a measure of the nation's economic growth that subtract the disamenities of an economy (costs of pollution, medical care, etc.) called the:
   a. Index of Sustainable Economic Welfare
 * b. Net Economic Welfare
   c. Gross Domestic Product
   d. Dow Average
   e. Average Growth Indicator

28. This measure of economic progress, created by economist Herman Daly includes such factors as the cost of pollution, cropland and wetland losses, costs of car accidents and a host of other factors:
 * a. Index of Sustainable Economic Welfare
   b. Net Economic Welfare
   c. Gross Domestic Product
   d. Dow Average
   e. Average Growth Indicator

29. The GNP continued to rise between 1980 and 1990 but the Index of Sustainable Economic Welfare declined. This decline is largely attributed to the:
   a. increase in personal spending.
   b. increase in pollution controls.
 * c. rapid deterioration of the environment.
   d. new market opportunities and investments.
   e. illegal immigration.

30. Gold mining in several African countries, provide little economic benefit to the miners and their local economies, whereas the middle men and others reap huge profits. Overpopulation and environmental degradation are fostered by the _____ of the miners.
   a. conditions
   b. attitude
 * c. economic exploitation
   d. work
   e. union organization

31. Which of the following responses is a characteristic of a sustainable economy?
   a. Uses resources efficiently.
   b. Restores damaged ecosystems.
   c. Promotes regional self-reliance.
   d. Relies on appropriate technology.
 * e. all of the above.

32. Green taxes on undesirable products or activities:
 * a. discourage production and use of unsustainable products or services.
   b. have no positive effects.
   c. are a threat to a democracy.
   d. are unfair to honest businesses.
   e. are unconstitutional at this time.

33. The gas-guzzler tax would be a good _____ to encourage a sustainable economy.
   a. economic incentive
 * b. economic disincentive
   c. rebate
   d. premium
   e. refund

34. _____ and _____ offer a 5% tax credit for companies that invest in recycling equipment.
   a. Iowa and Illinois
   b. Texas and Oregon
 * c. Colorado and Wisconsin
   d. Louisiana and Georgia
   e. Florida and Kansas

35. Trade and interdependence (global markets) allow human populations to:
   a. Flourish and reach their full economic potential.
   b. increase the GNP and prosper.
   c. exploit resources better for growth.
   d. foster local sustainability.
 * e. grow beyond the local and regional carrying capacity and increase environmental damage.

36. The Valdez Principles are:
    a. regulations to prevent future oil spills.
*   b. guidelines for environmentally responsible corporate conduct.
    c. are part of the latest oil pollution treaty.
    d. were created by the Earth Summit.
    e. EPA mandates to industry.

37. Environmentally friendly goods and services are called:
*   a. green products
    b. tree-hugger paraphernalia
    c. expendable resources
    d. recyclables
    e. organics

38. Processes that produce little, if any, pollution, use local resources and are efficient are called:
    a. green processes
*   b. appropriate technologies
    c. Green Seal technologies
    d. new age technologies
    e. whole earth technologies

39. Which of the following responses is a characteristic of appropriate technology?
    a. Machines are small to medium size.
    b. Production is decentralized.
    c. Products are durable.
    d. Products are generally for local consumption.
*   e. all of the above

40. The dollar value of goods per hour of paid employment is called the:
    a. net profit
    b. gross profit
    c. discount rate
    d. return on investment
*   e. productivity

257

27
Sustainable Economic Development: Challenges Facing the Developing Nations

Chapter Outline

Conventional Economic Development Strategies and Their Impacts
What Is Wrong with Western Development Assistance
Who's Financing International Development

Sustainable Economic Development Strategies
Employing Appropriate Technology
Creating Environmentally Compatible Systems of Production
Tapping Local Expertise and Encouraging Participation
Promoting Flexible Strategies
Improving the Status and Expanding the Role of Women
Preserving Natural Systems and Their Services
Improving the Productivity of Existing Lands

Overcoming Attitudinal and Economic Barriers
Attitudinal Barriers
Economic Barriers

Key Terms

sustainable development
attitudinal barriers
inappropriate technologies
production systems

appropriate technologies
economic barriers
international development

natural systems
developing countries
economic development

Objectives

1. List some of the problems with conventional economic development in developing countries.
2. Describe the goals of sustainable economic development.
3. Discuss some sustainable development strategies for developing nations.
4. Define and give an example of the following term: "appropriate technologies".
5. List the reasons why local expertise and participation should be considered in any development program.
6. Discuss the important role of women in developing countries and why they have a central role in sustainable activities
7. Summarize the attitudinal and economic barriers to sustainability.

Lecture Outline

I. Conventional Economic Development Strategies and Their Impacts - Population growth, political corruption, inter- and intranational conflicts, and industrial and governmental exploitation and intervention are some of the sources of economic and environmental problems in developing nations
   A. What is Wrong with Western Development Assistance
      1. International development assistance, though well-intentioned, has exacerbated many of the problems it sought to address.
      2. Some development projects have actually worked at cross purposes with others, each canceling the other's benefits.
      3. International development efforts which do not take into account the needs of the people and their particular ecosystem can undermine the establishment of sustainable practices and cultures.
   B. Who's Financing International Development?
      1. Financial backing for these projects comes primarily from multilateral development and private commercial banks, development agencies of developed nations, and private foundations.

II. Sustainable Economic Development Strategies - A sustainable economy will operate indefinitely within the limits imposed by nature; to achieve it will require population control, value shifts, and political restructuring.
   A. Employing Appropriate Technology
      1. Small scale, environmentally- and socially-compatible and responsive technology is appropriate technology. This is needed if developing nations are to avoid repeating the mistakes of the industrial nations.
   B. Creating Environmentally Compatible Systems of Production
      1. Sustainable development will be development geared to the human needs and environmental conditions of the country to which it is applied.
   C. Tapping Local Expertise and Encouraging Participation
      1. The knowledge and expertise of people indigenous to the region in need of assistance must be sought and utilized in making development decisions.
   D. Promoting Flexible Strategies

1. The traditional top-down approach to development must give way to a more flexible and humane bottom-up style initiated by local people perhaps in cooperation with NGOs.
   E. Improving the Status and Expanding the Role of Women
   1. Sustainable development is bound to fail if the needs and perspectives of women are not given equal status with those of men.
   F. Preserving Natural Systems and Their Services
   1. While sometimes initially expensive, efforts to protect and restore natural systems are integral to long-term economic and ecological health in developing countries.
   G. Improving the Productivity of Existing Lands
   1. Better land management practices can help reduce pressures for development of relatively undisturbed areas.

III. Overcoming Attitudinal and Economic Barriers
   A. Attitudinal Barriers
   1. A sustainable society will be one which adopts a worldview recognizing the necessity of efficiency, environmental protection, and individual responsibility.
   B. Economic Barriers
   1. Demilitarization, international debt reduction, and fairer distribution of resources are steps the wealthy nations can take to help foster sustainable development of less-developed nations.

Suggestions for Presenting the Chapter

- Instructors should stress that Western ideas, culture and technologies may not be appropriate or sustainable in developing nations.
- The idea of sustainable local development is a prevalent theme in this chapter. Sustainable local and regional development is essential in both developing and developed nations.
- There are many anthropological videos which illustrate that local cultural adaptations are sustainable. Videos can be viewed during lecture or assigned for later viewing.
- Investigating local and regional organizations involved in international development/aid projects is a good way to stimulate student awareness. Speakers are often available from these organizations.
- Recommended web sites:

   World Business Council for Sustainable Development:http://www.wbcsd.ch/aboutus.htm#top
   United Nations Commission on Sustainable Development: http://www.un.org/esa/sustdev
   International Institute for Environment and Development: http://www.iied.org
   International Development Research Center: _http://www.idrc.ca
   World Health Organization/Health and Environment in Sustainable Development:
   http://www.who.int/environmental_information/Information_resources/htmdocs/execsum.htm

True/False Questions

1. ___ In developing countries, people spend 40% of their income on food.
2. ___ Sustainable development is both people-centered and conservation-based.

3. ___ Overpopulation, political corruption, misguided development efforts and war are factors that may exacerbate environmental problems in developing countries.
4. ___ Most development strategies have attempted to impose Western ways on developing nations.
5. ___ Huge dams built for hydropower may flood productive farmland and displace farmers.
6. ___ Western development often undermines sustainable practices and sustainable cultures.
7. ___ The World Bank is a multilateral development bank.
8. ___ To create a sustainable future, developing nations must slow population growth.
9. ___ Appropriate technology refers to an assortment of environmentally compatible technologies.
10.___ Nongovernmental organizations are private groups involved in sustainable development.
11.___ Bhutan is a country nestled in the Swiss Alps.
12.___ In most developing countries men collect firewood, maintain the gardens and care for the children.
13.___ Environmental protection is a luxury that should be addressed after people of developing countries have achieved prosperity.
14.___ Environmental protection is an indulgence that can only be afforded by the wealthy in developing countries.
15.___ Developing nations will not need much money to develop in a sustainable fashion.
16.___ Energy efficiency projects constitute less than 1% of all international aid to developing nations.
17.___ Environmental decay drives many people into poverty in developing nations.
18.___ One economic barrier to environmental protection in developing nations might be there large expenditures for military protection.
19.___ Wealthy nations can improve the economic conditions of developing nations by ending their exploitation of natural resources.
20.___ Debt-for-nature swaps is one way to relieve the burden of international debt many developing nations carry.

True/False Key:

1. T 2. T 3. T 4. T 5. T 6. T 7. T 8. T 9. T 10. T 11. F 12. F 13. F 14. F 15. F 16. F 17. T 18. T 19. T 20. T

Fill-in-the-Blank Questions

1. In Central America, chronic and acute _____ poisoning are some of the region's most serious problems.
2. The disruption of local cultures during the development process is known as _____ erosion.

3. Huge dams built for _____ often flood productive farmland and displace farmers.
4. Today, over _____ of the irrigated cropland in developing countries suffers from salinization.
5. _____ technologies are reliant on locally available resources and knowledge for repair and service.
6. Tractors and diesel-powered water pumps are often _____ technologies for developing nations.
7. _____ box cookers are suitable in regions where firewood is scarce and sunlight abundant.
8. _____ are far cheaper than centralized, coal-fired power plants in rural regions of developing countries.
9. Gandhi summed up the challenge of development when he said, "the poor of the world cannot be helped by mass production, only by _____ of the masses".
10. Western-style development has been driven by _____ ethics.
11. Farming practices that may be suitable in Iowa often fail miserably in the _____.
12. Convinced of the superiority of Western ways, many experts overlook the knowledge held by _____ peoples.
13. The Lancandons of Mexico practiced agroforestry and grew as many as ___ different crops in the rain forest.
14. Development is unlikely to be sustained unless the needs of _____ are identified and local residents support the project.
15. Developers might be advised to find projects that promote the good of the entire _____, rather than enterprises that promote private ownership and competition.
16. Large bureaucracies lack the _____ to respond to problems because they tend to be organized from the top-down.
17. _____ play an important role in shaping a country's future but are often relegated to an inferior position.
18. In most developing countries, women are important managers of _____ resources.
19. Sustainable development clearly requires steps to _____ the status of women.
20. Environmental protection and efficiency are often seen as _____ in developing countries.

Fill-in Key:

1. pesticide
2. cultural
3. hydropower
4. half
5. appropriate
6. inappropriate
7. solar
8. photovoltaic
9. production
10. frontier
11. tropics

12. indigenous
13. 80
14. people
15. community
16. flexibility
17. women
18. natural
19. improve
20. luxuries

Multiple Choice Questions

1. Which of the following responses is a problem with western style development in developing nations?
    a. Western help tends to centralize wealth and destroy sustainable lifestyles.
    b. Western development radically reshapes the environment.
    c. Western development encourages depletion of natural resources.
    d. Western development promotes a costly dependency on Western economies.
    * e. all of the above

2. The harmful influence of Western development on sustainable practices and sustainable cultures is called:
    a. assimilation
    * b. cultural erosion
    c. ethnocentrism
    d. exploitation
    e. dominance

3. Financing for international development comes from:
    a. multilateral development banks
    b. private commercial banks
    c. development agencies of the industrial nations
    d. private foundations
    * e. all of the above

4. To create a sustainable future developing nations must:
    a. slow population growth.
    b. become more self-sufficient.
    c. improve education and health care.
    d. build sustainable systems of agriculture and commerce.
    * e. all of the above

5. Solar-powered pumps, photovoltaic cells and solar box cookers are examples of:
    a. new age technologies
    b. expensive and unfeasible technology
    * c. appropriate technology
    d. camping equipment
    e. unsustainable development

6. The Lancandons are:
    a. an extinct tribe from the Amazon region.
    b. a group of Irish farmers.
    * c. a Mexican tribe descended from the Mayans.
    d. wheat farmers in Alberta.
    e. the first settlers in Pennsylvania.

7. Creating flexible management structures that are more responsive to emergent needs in developing countries may require removing international development agencies and banks and replacing them with:
    a. government agencies
    b. unions
    * c. nongovernmental organizations
    d. international aid agencies like Red Cross
    e. private banking organizations

263

8. Working Women's Forum is an organization that seeks to:
   a. organize women into labor unions.
   b. provide credit to rural women to help them start small businesses.
   c. help women acquire governmental aid for prenatal and child care, immunization, and family planning.
   d. a and b only
   * e. b and c only

9. To be sustainable, social and economic development strategies must _____ natural systems, which provide free services such as flood control and water purification.
   a. exploit
   * b. protect
   c. harvest
   d. economically dominate
   e. change

10. The World Bank's new guidelines for international development:
   a. encourage forest exploitation and timber harvest.
   b. foster Western style development projects in lieu of sustainable local solutions.
   * c. strictly prohibits supporting projects that will destroy wildlands of special concern, including wetlands and rainforests.
   d. encourage oil and mineral exploration and development in rainforests around the world.
   e. eliminate the need for nongoverrnmental organizations.

11. Today energy efficiency projects constitute less than __% of all international aid.
   * a. 1
   b. 10
   c. 15
   d. 20
   e. 25

12. One study showed that if the entire world population lived like the North Americans, resources would last about:
   * a. 3 months
   b. 3 years
   c. 3 decades
   d. 3 centuries
   e. 3 millennia

13. The country of Bhutan reacted to exploitation of its forest resources by:
   a. encouraging export of wood.
   b. passing legislation to limit the exploitation.
   * c. nationalizing most logging operations.
   d. expanding logging production.
   e. increasing import of wood products.

14. Appropriate technologies depend more on _____ than on machines and fossil fuel energy.
   a. exploitation of resources
   b. ancient technologies
   c. tribal customs
   * d. human labor
   e. computers

15. The use of diesel powered tractors and pumps may be inappropriate technologies for developing nations because they:
   a. displace farm workers.
   b. are expensive to operate.
   c. are expensive to repair.
   d. a and b only
   * e. a, b and c

16. The solar box cooker uses sunlight to cook meals, is inexpensive and can:
   a. reduce health problems due to smoke inhalation.
   b. reduce deforestation.
   c. be suitable in regions where firewood is scarce and sunlight is abundant.
   d. a and b only
   * e. a, b, and c

17. Which of the following responses is a characteristic of Western-style development?
*   a. Promoting individual success and gain.
    b. Rural people working together for the good of the entire citizenry.
    c. Dependence on locally produced goods and services.
    d. Respecting local knowledge of the environment and cultural traditions.
    e. Viewing local cultures as the key to achieving a sustainable development strategy.

18. Which of the following responses is a problem of traditional development strategies?
    a. Bureaucracies tend to manage large projects great distances from their headquarters.
    b. Large bureaucracies tend to be inflexible and may not overcome obstacles easily.
    c. Local knowledge is often ignored in face of Western expertise.
    d. Development projects may be culturally insensitive.
*   e. all of the above

19. Sustainable development clearly requires steps to improve the status of:
    a. farm workers.
    b. laborers
    c. carpenters
*   d. women
    e. multinational corporations

20. Which of the following responses is a potential barrier to sustainable development?
    a. Environmental protection is a luxury.
    b. Efficiency is a luxury.
    c. Huge sums of money will be required by developing nations.
    d. a and b only
*   e. a, b and c

28

Law, Government, and Society

Chapter Outline

The Role of Government in Environmental Protection
Forms of Government
Government Policies and Sustainability

Political Decision Making: The Players and the Process
Government Officials
The Public
Special Interest Groups
Environmental Groups

Environment and Law: Creating a Sustainable Future
Evolution of U.S. Environmental Law
The National Environmental Policy Act
The Environmental Protection Agency
Principles of Environmental Law
Resolving Environmental Disputes Out of Court

Creating Governments That Foster Sustainability
Creating Governments with Vision
Ending Our Obsession with Growth

Global Government: Toward a Sustainable World Community
Regional and Global Alliances
Strengthening International Government

Key Terms

government	democratic nations	communist nations
tax credit	laws	public policy
theory of public choice	political action committees	ecotage
civil disobedience	statutory law	common law
plaintiff	defendant	balance principle
nuisance	negligence	concept of knowing
burden of proof	mediation	statute of limitations

crisis politics   proactive laws   Green Party
national security   world government

Objectives

1. Compare the forms of government and their roles in environmental protection.
2. Discuss how democratic governments regulate activities such as environmental protection.
3. Summarize the roles that government officials, the public, special interest, and environmental groups play in environmental politics.
4. Outline the development of U.S. environmental law at the federal, state and local levels.
5. Discuss the major legal principles which form the basis for environmental law.
6. List some recommendations for creating governments that foster sustainability.
7. Suggest some reasons that global government or global organizations can be useful in obtaining global sustainability.

Lecture Outline

I. The Role of Government in Environmental Protection
    A. Forms of Government - In general, free market economies predominate in democratic nations, while command economies are found in nations with communist or socialist governments.
    B. Government Policies and Sustainability
        1. Governments regulate activities and protect the environment through taxes, expenditures, and regulations.
        2. Taxes help regulate behavior and raise funds for government expenditures, such as pollution-control project grants and procurements programs; in poor countries, such funds are scarce.
        3. Regulations take the form of federal laws or agency-promulgated regulations which govern specific activities.

II. Political Decision Making: The Players and the Process
    A. Government Officials
        1. Government officials have the most power in communist nations, but often have final say on certain policies in any type of government.
    B. The Public
        1. Voters influence policy in democratic nations by selecting representatives and by communicating their priorities and concerns to those in office. Even communist governments are somewhat responsive to public sentiments and pressures.
        2. Special Interest Groups
            a. Special interest groups, such as automakers and environmental organizations, can exert leveraged and sometimes disproportionate influence on policymakers, through PAC's and lobbying.
            b. Environmental groups also affect public policy through public displays, educational materials, awareness-raising activities, pollution monitoring, legal challenges, protests, and interventions.

III. Environmental Law: Creating a Sustainable Future
   A. Evolution of U.S. Environmental Law
      1. State and federal environmental laws gradually evolved from scattered local ordinances which limited activities of a few for the good of all.
      2. Conflicts between neighboring municipalities necessitated pollution controls at the state level.
      3. Because pollution crosses state lines, interstate conflicts arose; in response, environmental legislation was enacted at the federal level.
      4. The federal government is best suited to regulate in situations requiring uniform standards and large expenditures.
   B. National Environmental Policy Act
      1. NEPA is a landmark piece of U.S. environmental legislation which introduced requirements for environmental impact statements and set a goal of minimum environmental impact for all federal projects.
   C. Environmental Protection Agency
      1. Founded in 1970, the EPA manages many of the major environmental laws written by Congress and conducts research, provides grants, and otherwise influences policy and action related to environmental issues.
   D. Principles of Environmental Law
      1. Statutory
         a. Statutory laws state broad principles, for which specific standards are set by EPA or other agencies.
      2. Common-law
         a. Common law is a body of unwritten rules and principles derived from countless legal precedents.
         b. Through common law, competing interests are weighed and, ideally, fairly protected.
         c. Most common law cases are decided on the basis of two legal principles: nuisance and negligence.
      3. Problems with Environmental Lawsuits
         a. Burdens of proof, statutes of limitations, and out-of-court settlements have all presented legal problems to those trying to settle environmental lawsuits.
   E. Resolving Environmental Disputes Out of Court
      1. Mediation or dispute resolution is increasingly used to settle environmental disputes out of court; it is less costly, less time-consuming, and less adversarial.

IV. Creating Governments That Foster Sustainability
   A. Creating Government with Vision - Lack of consensus or agreement about goals and priorities sometimes prevents positive action; a stronger, more coherent vision is needed to overcome this obstacle.
      1. Increasing Public Awareness Through Research and Education
         a. By reducing empirical uncertainty, research can help establish goals for long-range planning.
         b. Education can help galvanize the public in commitment to solving environmental problems.
      2. Getting Beyond Crisis Management
         a. Reacting to urgent, immediately pressing problems, rather than proacting to deal with important long-term problems, is characteristic of crisis politics.

3. Getting Beyond Limited Planning Horizons
    a. The planning horizon of political decision makers is unduly shortened by budget periods, turns in office, and reelection concerns; most environmental protection measures require longer planning horizons and payback periods.
    b. A sustainable society must redefine its primary goals in view of long-term considerations.
4. Becoming Proactive
    a. Reactive governments primarily address urgent, immediate issues with proposals for remedial action.
    b. Proactive government takes a long-term outlook and aims to prevent, rather than solve, problems.
    c. Most governments mix reactive and proactive policies and approaches, but reaction predominates in modern political systems.
B. Ending Our Obsession with Growth
    1. By replacing current legislation which emphasizes growth with policies promoting sustainability, Congress would be sending an important message to the American people and the world.
    2. Reducing Exploitation and Promoting Self-Reliance
        a. Sustainability Through Land-Use Planning
            i. Poor or inadequate land-use planning puts land to unsustainable uses and can be ruinous.
            ii. Proper land-use planning manages resources for maximum sustainability and long-term productivity.
            iii. Zoning is the main tool of land-use planners; it can be used in conjunction with differential tax assessment laws and purchase of development rights to protect resources.
    3. Models of Sustainable Development
        a. The Greens are a political party actively pushing for a proactive, long-range approach to government; their goal is creation of a sustainable society.

V. Global Government: Toward a Sustainable World Community - A new and appropriate notion of national security will necessarily be based on protection of the environment from internal and external threats.
A. Regional and Global Alliances
    1. The Climate Convention
        a. Intended to slow the rate of global warming, 154 nations have signed this agreement to limit greenhouse gas emissions.
    2. The Biodiversity Convention
        a. Aimed at conservation and preservation of biological resources worldwide, then President Bush refused to sign this agreement.
    3. Agenda 21
        a. Despite its weaknesses, this document is a testimonial to the success of the Earth Summit in achieving international cooperation towards sustainable development.
    4. Forest Principles
        a. The sovereignty of nations to exploit their forests overrode concern for global ecological health in drafting this set of principles.
    5. Rio Declaration

a. This general statement of principles from the Earth Summit makes a number of recommendations for sustainability and begins to address the issues of international obligation and responsibility.
B. Strengthening International Government
1. Strengthening the U.N.'s Role in Sustainable Development International cooperation and strengthened commitment are necessary to achieve this goal.
2. Creating a World Government
a. Though requiring an entirely new global perspective, this may be necessary to effectively address global environmental and social problems in the long term.

Suggestions for Presenting the Chapter

- Instructors should foster an awareness of the current environmental regulatory structure in the United States.
- A trip to a local industry and tour with the person in charge of environmental compliance is recommended. The impact of current regulations on the environment should be emphasized.
- Time should be spent looking at the global impacts of government laws and regulations. A discussion about the effectiveness of the United Nations, the Earth Summit and a proposed world government are good topics.
- Have the students examine what environmental laws directly effect their lives or the operation of your educational institution. Does your institution have an environmental policy? Is your institution in compliance with current regulations? How does your institution handle compliance with environmental regulations?
- Recommended web sites:

    Earth Justice Legal Defense Fund: http://www.earthjustice.org/news/
    Friends of the Earth: http://www.foe.org/
    The Center for International Environmental Law: http://www.ciel.org/
    Community and Environmental Defense Services: http://www.ceds.org/
    Honor the Earth: http://www.honorearth.com/

True/False Questions

1. ___ A tax credit is a dollar amount of the purchase price that a person can deduct from income or corporate tax.
2. ___ President Reagan openly opposed the environmental protection goals of previous administrations.
3. ___ Lobbyists are individuals who work to convince legislators of the merits of their particular views.
4. ___ The Double-C/Double-P game is a board game resembling *Monopoly*.
5. ___ Ecotage is sabotage in the name of the environment.
6. ___ The Nature Conservancy is an environmental organization that purchases land for environmental protection.
7. ___ The Wise Use Movement is strongly supporting environmental protection efforts.

8. ___ Environmental impact statements are required on federal lands as a result of the National Environmental Policy Act.
9. ___ The U.S. Department of Energy manages and enforces most of the country's environmental laws.
10. ___ Statutory laws arise in legislative bodies such as Congress.
11. ___ Common law is based on proper or reasonable behavior.
12. ___ A public nuisance is an activity that interferes with the rights of few people.
13. ___ Class action suits are filed on behalf of many people and seek remedy for damage caused to the entire group.
14. ___ A person is negligent if he or she acts in an unreasonable manner and if these actions cause personal or property damage.
15. ___ Industrial ecology is a government mandated attempt to cleanup dirty industries.
16. ___ Mediation is much more costly and more time consuming to litigants settling cases out of court.
17. ___ The statute of limitations limits the length of time within a person can sue or be prosecuted.
18. ___ Research and education are essential to raise the level of awareness of environmental problems and sustainable solutions among the public and elected officials.
19. ___ The Green Party in Germany has been instrumental in encouraging sustainable practices.
20. ___ The National Appliance Energy Conservation Act establishes efficiency standards for appliances.

True/False Key:

1. T 2. T 3. T 4. F 5. T 6. T 7. F 8. T 9. T 10. T 11. T 12. F 13. T 14. T 15. F 16. F 17. T 18. T 19. T 20. T

Fill-in-the-Blank Questions

1. Sustainability relies in large part on _____ laws that attempt to prevent problems.
2. Using a neutral party to resolve lawsuits is called _____.
3. The length of time that a person can sue or be prosecuted after a violation occurs is known as the statute of _____.
4. The Price-Anderson Act frees utility companies from financial liability incurred by _____ power plant accidents.
5. _____ occurs when a person acts in an unreasonable manner and if their actions cause personal or property damage.
6. The party that files a lawsuit is the _____.
7. A public _____ is an activity that harms or interferes with the rights of the general public.
8. _____ laws generally establish broad goals and are passed by legislative bodies.
9. The EPA was founded by an executive order of President _____.
10. The National Environmental Policy Act was enacted in _____.

11. The environmental impact of projects on federal lands must be detailed in the _____ impact statement required by the National Environmental Policy Act.
12. U.S. environmental law began _____ and evolved to higher levels of government.
13. The Wise Use Movement is heavily funded by _____ companies, mining companies, oil and coal companies, cattle ranchers and other industries opposing environmental regulations.
14. _____ are people working to convince legislators of the merits of their particular political interests.
15. Double-C/Double-P stands for "_____ the Costs and Privatize the Profits".
16. Democratic governments regulate activities through _____ policy, direct financial support and laws.
17. A tax _____ is the amount of the purchase price that a individual or business can deduct from income or corporate taxes.
18. _____ law is a body of unwritten rules and principles derived from thousands of years of legal decisions.
19. _____ probably has the highest environmental standards in the world.
20. The _____ Convention call on nations to take steps to protect species.

Fill-in Key:

1. proactive
2. mediation
3. limitations
4. nuclear
5. negligence
6. plaintiff
7. nuisance
8. statutory
9. Nixon
10. 1969
11. environmental
12. locally
13. timber
14. lobbyists
15. commonize
16. tax
17. credit
18. Common
19. German
20. Biodiversity

Multiple Choice Questions

1. The dollar amount of the purchase price that a person or company can deduct from income or corporate tax is called a:
   a. rebate
   * b. tax credit
   c. subsidy
   d. refund
   e. surcharge

2. A tax credit for energy conservation passes a portion of the cost of the improvements to:
   a. private individuals
   * b. taxpayers
   c. energy companies
   d. government agencies
   e. corporations

3. Democratic governments regulate environmental protection by:
   a. tax policy
   b. direct financial support
   c. laws and regulations
   d. a and b only
   * e. a, b. and c

4. The theory of public choice states that politicians act in ways that:
   a. minimize bad publicity.
   * b. maximize their chances of reelection.
   c. attract PAC money to their campaigns.
   d. alienate the public.
   e. insure corruption in each election.

5. Commonize the Costs and Privatize the Profits is the:
   a. common technique used to win reelection.
   b. economic theory of the 1940s.
   * c. Double-C/Double-P game.
   d. way PACs finance campaigns.
   e. theory of public choice.

6. Which of the following is not an environmental group?
   a. Earth First
   b. Greenpeace
   c. Sierra Club
   * d. Wise Use Movement
   f. Environmental Defense Fund

7. This U.S. law requires an environmental impact statement be written for federally controlled or funded projects:
   a. Resource Conservation and Recovery Act
   b. Clean Water Act
   * c. National Environmental Policy Act
   d. Occupational Safety and Health Act
   e. Oil Pollution Act

8. An Environmental Impact Statement must describe:
   a. What the project is.
   b. The need for the project.
   c. Its short and long term environmental impact.
   d. Proposals to minimize the impact, including alternatives to the project.
   * e. all of the above

9. This state passed an Environmental Quality Act in 1970 that requires Environmental Impact Statements for all projects, private and public if they will affect the environment.
   a. Oregon
   * b. California
   c. Idaho
   d. Washington
   e. Wisconsin

10. The U.S. Environmental Protection Agency was created by:
    a. an act of Congress.
    b. National Environmental Policy Act
    * c. executive order of President Nixon
    d. Clean Air Act
    e. Congressional order

11. Laws written and agreed upon by legislative bodies are called:
    a. common law
    * b. statutory law
    c. tort law
    d. criminal law
    e. case law

12. A body of unwritten rules and principles derived from thousands of years of legal decisions is:
    * a. common law
    b. statutory law
    c. tort law
    d . criminal law
    e. case law

13. The most common ground for legal action in the field of environmental common law is
    a. negligence
    b. precedence
    * c. nuisance
    d. damages
    e. suffering

14. An activity that harms or interferes with the rights of the general public is a:
    a. private nuisance
    * b. public nuisance
    c. hazard
    d. liability
    e. danger

15. Lawsuits filed on behalf of many people and seek remedy for damage caused to the entire group by a nuisance is a:
    * a. class action suit
    b. criminal case
    c. negligence suit
    d. grievance
    e. lien

16. The production of goods and services in ways that protect and enhance the environment is called:
    a. industrial management
    * b. industrial ecology
    c. the Green Revolution
    d. restoration ecology
    e. wise use

17. An unreasonable act that causes personal or property damage is called a:
    * a. negligence
    b. nuisance
    c. violation
    d. misdemeanor
    e. felony

18. The Price-Anderson Act is a law eliminating liability incurred by:
    a. private citizens
    b. water polluters
    * c. nuclear power plants
    d. airlines
    e. oil companies

19. The length of time that a suit can be brought after a particular event is called the:
    a. timeline
    b. terminal time
    * c. statute of limitations
    d. time limit
    e. compensation time

20. The process of settling disputes out of court or by dispute resolution is called:
    a. remediation
    * b. mediation
    c. mitigation
    d. amelioration
    e. accommodation

21. The management of immediate problems at the expense of long-range problems is called:
    a. planning
 * b. crisis management
    c. sustainable management
    d. poor planning
    e. disorganization

22. Raising the level of awareness of environmental problems and sustainable solutions among the public and elected officials requires:
    a. research
    b. education
    c. media
    d. environmental groups
 * e. all of the above

23. The Critical Trends Assessment Act was introduced to Congress in 1985 by:
 * a. Senator Al Gore
    b. Senator Edward Kennedy
    c. Senator Tom Harkin
    d. Senator Gaylord Nelson
    e. Senator Hugh Heflin

24. This organization established the World Commission on Environment and Development:
    a. U.S. Congress
    b. Environmental Defense Fund
 * c. United Nations
    d. North Atlantic Treaty Organization
    e. World Bank

25. Which of the following acts of Congress established energy efficiency standards for appliances?
    a. Resource Conservation and Recovery Act
    b. National Environmental Policy Act
 * c. National Appliance Energy Conservation Act
    d. Office of Technology Assessment
    e. Occupational Safety and Health Act

26. Which of the following is a recommendation encouraging sustainable government?
    a. Repeal or modify existing laws that hinder sustainability.
    b. Pass a national Sustainable Futures Act
    c. Establish offices of critical trends analysis.
    d. Adopt sustainable land-use planning at the state and national level.
 * e. all of the above

27. This political group has been active in Germany and other countries promoting social justice and a healthy environment:
    a. Wise Use Movement
 * b. Green Party
    c. World Bank
    d. World Wildlife Fund
    e. libertarian party

28. The International Whaling Commission:
    a. promotes commercial exploitation of whales.
 * b. sets quotas on whale kills and enacts outright bans on whaling.
    c. is part of the United Nations.
    d. has a large fleet of enforcement vessels.
    e. is empowered to imprison violators of its decrees.

29. Which of the following programs is funded by the United Nations?
    a. Fund for Population Activities
    b. Food and Agricultural Organization
    c. United Nations Environment Programme
    d. United Nations Development Program
 * e. all of the above

30. The most comprehensive document to emerge from the Earth Summit was:
    a. The Climate Convention
    b. The Biodiversity Convention
 * c. Agenda 21
    d. The Helsinki Accord
    e. Authoritative Statement of Principles on the World's Forests.